For over 14 years, we have ~~been engaged in~~
operations that have demonstrated our unwavering resolve to meet and destroy
our enemies on the battlefield. In the face of challenges, we preserved our
reputation as the most powerful, professional, and respected Army in the
world. Although our Army is a combat-seasoned force, we must not be lauded
into a false sense of pride or confidence that our enemies can never defeat us.
We must remain vigilant and committed more than ever to building a leaner,
more agile, adaptive force whose ranks are made up of critical thinking
Soldiers and leaders who will win on any battlefield.

The meaning "Backbone of the Army," resonates across our formations
to remind us that for over 239 years, the NCO Corps has carried the
responsibility of training, caring for, and developing Soldiers. As the Army
continues to evolve and develop a force for 2025 and beyond, the NCO of
2020 will form an NCO Corps, grounded in the heritage, values, and tradition,
which embodies the warrior ethos; values perpetual learning; and is capable of
leading, training, and motivating Soldiers in a complex world. We must invest
in our NCO Corps through rigor and discipline developing resilience, agility,
and adaptability.

This guide is a tool to assist you in establishing standards,
understanding your duties and responsibilities, and arming you with
knowledge to lead your team, section, squad, or platoon. You are charged and
challenged to take ownership of the knowledge contained in this guide and
apply its principles to affirm your character, competence, confidence, ethics,
and values based on untarnished action and conduct. Develop your Soldiers
applying the art and science of leading while carrying the mantle of a servant
leader. Your Soldiers and organizations, the Army, and our Nation have
placed great trust in you and expect your best. You are their Trusted
Professional. Army Strong!

DANIEL A. DAILEY
Sergeant Major of the Army

RAYMOND T. ODIERNO
General, U.S. Army
Chief of Staff

*TC 7-22.7 (FM 7-22.7)

Training Circular
No 7-22.7

Headquarters
Department of the Army
Washington, DC, 7 April 2015

Noncommissioned Officer Guide

Contents

Distribution Restriction: Approved for public release, distribution is unlimited.
*This publication supersedes FM 7-22.7, dated 23 December 2002.

FIGURES

Preface

This Department of the Army Training Circular (TC) is dedicated to the men and women of the U.S. Army NCO Corps, who have made the ultimate sacrifice and to NCOs presently serving in the Active Army, the Army National Guard, and the Army Reserve. Your Soldiers will seek your guidance; trust you to train them, and to develop them to win the nation's wars. As the standard-bearer, instill pride and strive to live the Army Values. You are "The Backbone of the Army." [1]

Scope. This TC provides the Army's NCOs a guide for leading, supervising, and caring for Soldiers. While this guide is not all-inclusive, nor is it intended as a stand-alone document, it provides NCOs a quick and ready reference to refresh and develop leadership traits.

Applicability. This TC provides critical information for the success of today's NCO and is intended for use by all the NCOs in the Active Army Component, the Army National Guard, and the Army Reserve Component.

Proponent and Exception Authority Statement. The proponent for this TC is the United States Army Sergeants Major Academy.

Interim Changes. None

Suggested Improvements. Send comments and recommendations on a DA Form 2028 (Recommended Changes to Publications and Blank Forms) to Deputy Director, DOT, United States Army Sergeants Major Academy, and ATTN: ATSS-DAE, Fort Bliss, Texas 79918-8002.

Purpose: To use as a guide to develop an innovative, competent professional NCO. NCOs must learn to analyze and evaluate the operational environment to create and apply an understanding of the changing world that confronts them.

Distribution. Unlimited.

---

[1] NCO Creed. http://www.army.mil/values/nco.html

CHARGE of the NONCOMMISSIONED OFFICER

𝕴 will discharge carefully and diligently the duties of the grade to which I have been promoted and uphold the traditions and standards of the Army.

𝕴 understand that Soldiers of lesser rank are required to obey my lawful orders. Accordingly, I accept responsibility for their actions. As a noncommissioned officer, I accept the charge to observe and follow the orders and directions given by supervisors acting according to the laws, articles and rules governing the discipline of the Army. I will correct conditions detrimental to the readiness thereof. In so doing, I will fulfill my greatest obligation as a leader and thereby confirm my status as a noncommissioned officer.

COMMAND SERGEANT MAJOR   NONCOMMISSIONED OFFICER

Figure 1. Charge of the Noncommissioned Officer

THE NONCOMMISSIONED OFFICER VISION

An NCO Corps, grounded in heritage, values and tradition, that embodies the warrior ethos; values perpetual learning; and is capable of leading, training and motivating soldiers.

We must always be an NCO Corps that
- Leads by Example
- Trains from Experience
- Maintains and Enforces Standards
- Takes care of Soldiers
- Adapts to a Changing World

Effectively Counsels and Mentors Subordinates
Maintains an Outstanding Personal Appearance
Disciplined Leaders Produce Disciplined Soldiers

SMA Jack L. Tilley
12th Sergeant Major of the Army

Figure 2. Noncommissioned Officer Vision

Introduction

NCOs conduct the daily operations of the Army. NCOs are relied on to execute complex tactical operations, make intent-driven decisions and operate in joint, interagency, and multinational environments. NCOs are responsible for maintaining and enforcing standards and a high degree of discipline. NCOs process Soldiers for enlistment, teach basic Soldier Skills, are accountable for the care of Soldiers, and set the example. NCOs are trainers, mentors, advisors, and communicators.

Every Soldier has a Sergeant, and every Soldier deserves a leader who is a capable trainer, is trustworthy, is genuinely concerned for their health and welfare, and develops them to be the leaders of tomorrow. The Soldier's Creed and Warrior Ethos are compelling obligations we expect our Soldiers to live by. Likewise, we expect our leaders to live by those obligations and those of the NCO Creed.

As the culture of the Army changes, we face tremendous challenges. How we communicate, use technology, increase resilience, sustain tactical and technical proficiency, and inculcate ourselves and our Soldiers on ethics and values are critical to maintaining an "Army Strong" force [2]. Soldiering is and has always been an affair of the heart. Leading is a privilege and an honorable profession. The two major responsibilities of leadership remain the accomplishment of the mission and the welfare of Soldiers. This guide is intended to enhance your ability to lead and arm you with tools such as self-awareness, self-discovery, how you lead, and how you develop your subordinates. It also ties into how competent and confident you are in your duties, responsibilities, and roles.

---

[2] Nickerson, Thomas COL, "The Making of Army Strong." November 8, 2006. The United States Army Home Page. http://www.army.mil/ (Accessed August 26, 2014).

CHAPTER 1

HISTORY OF THE NCO

Sergeant John Hill riding on Jumping Dan Ware, the finest jumping horse in the Infantry Stables.
Ft. Benning, Georgia (July 25, 1941) Photo #161-SC-41-1323 by the 161st Sig. Photo. Co

New Guinea. *Radio Operator, Cpl. John Robbins of Louisville, Nebraska, 41st Signal, 41st Inf. Div., operating his SCR 188 in a sandbagged hut at Station NYU. Dobodura, New Guinea. (9 May 43) Signal Corps Photo: GHQ SWPA SC 43 5901 (T/4 Harold Newman)*

SSG James L. Leach (detector dog handler) with "bomb" dog Jupiter (Brand Number 367B in left ear) used as team to check vehicles entering compound [military vehicles normally exempted from anti-terrorist check], 118th Military Police Company security at entrance to XVIII Airborne Corps Main Command Post, Rafha Airport, Northern Province, Saudi Arabia, 8 February 1991, XVIII Airborne Corps History Office photograph by SSG LaDona S. Kirkland, DS-F-163-02

Chapter 1

History of the NCO

1.0. Ref.

- ADP 6-22, Army Leadership
- ADRP 6-22, Army Leadership
- Arms, L.R., (2007). A history of the NCO. U.S. Army Sergeants Major Academy
- FM 6-22, Army Leadership: Competent, Confident, and Agile
- http://usacac.army.mil/core-functions/military-history

1.1. History.

a. The history of the NCO is an integral part of our Army and country's founding. As a corps, we originated with the establishment of the Continental Army of the American colonies on 14 June 1775. Colonial officers adopted the customs of the English military, where the colonial NCOs borrowed from Prussia, France, and some English traditions. Before Valley Forge, the practices and standards of the NCO were not uniformly enforced or formally written. In the winter of 1777-78, while encamped at Valley Forge, Pennsylvania, General George Washington acquired the talents of Prussian General Friedrich Wilhelm von Steuben, who wrote the "Regulations for the *ORDER* and *DISCIPLINE* of the Troops of the United States, what we commonly refer to as the "Blue Book."[3] The Blue Book codified the sphere of *RESPONSIBILITY* of each NCO, which included at the time, the ranks of corporal (CPL), Sergeant (SGT), First Sergeant (1SG), quartermaster Sergeant, and Sergeant Major (SGM).[4]

b. Rank Insignia - In 1821, the War Department made the first reference to NCO chevrons. A General Order directed that SGMs and quartermaster Sergeants wear a worsted

---

[3] L.R. Arms, *A Short History of the NCO*, U.S. Army Sergeants Major Academy, 1 November 1991.
[4] Ibid.

chevron on each arm above the elbow; SGTs and senior musicians, one on each arm below the elbow; and CPLs, one on the right arm above the elbow. This practice ended in 1829 but returned periodically and became a permanent part of the NCOs' uniform before the Civil War.[5]

c. In 1829, the Army published a manual titled *Abstract of Infantry Tactic*, which formalized the *ROLE* of the NCO. This established the first *TRAINING* program for NCOs to equip them with the necessary Soldiering skills of the time.[6] It outlined what SGM was responsible for teaching seasoned SGTs and CPLs and new SGTs and CPLs were trained by the 1SG (Figure 3).

Figure 3. The American Soldier-1836

---

[5] L.R. Arms, *A Short History of the NCO*, U.S. Army Sergeants Major Academy, 1 November 1991.

[6] U.S. Department of War, *Abstract of Infantry Tactics*, 2 March 1829

d. With the on-set of the Civil War (Figure 4), NCOs gained importance, where they led the front lines of fighting units and carried the identifying flag of their respective regiments. Improvements in technology during the Civil War, the telegraph and the railroad, provided venues to learn technical skills necessary for military operations and to earn higher pay.

Figure 4. Confederate and Union Soldiers

e. The Modern Rank Insignia – In 1902, the NCO symbol of rank, the chevron, rotated to what we now call point up and became smaller in dimension. Though many stories exist as to why the chevron's direction changed, the most probable reason was simply that it looked better. Clothing had become more form fitting, creating narrower sleeves; in fact, the 10-inch chevron of the 1880s would have wrapped completely around the sleeve of the1902 uniform.

f. By 1917 and our involvement in World War I (Figure 5), NCO roles and *DUTIES* were improved and expanded with the publication of the *Noncommissioned Officers' Manual.*[7] It described how frontier wars cultivated the NCO as a small unit leader. From World War II onward, the Army was provided time to prepare Soldiers for deployment and NCOs assumed the role of training troops. Since then, the NCO corps became professionalized at the inception of the NCO education system, which was designed to educate uniformly and train combat and leadership skills.

Figure 5. The American Soldier-1918

g. The rapid pace and acceptance of technology during the late 1930s caused the Army to create special "technician" ranks in grades 3, 4, & 5 (CPL, SGT & Staff Sergeant (SSG)), with chevrons marked with a "T."[8] This led to an increase in promotions among technical personnel. The technician ranks ended in 1948, but they later reappeared as 'specialists' in 1955.[9]

---

[7] James A. Moss, *Noncommissioned Officers' Manual.* 1917.

[8] L.R. Arms, *A Short History of the NCO*, U.S. Army Sergeants Major Academy, 1 November 1991.

[9] Ibid.

h. On 17 October 1949, the doors opened for the first Noncommissioned Officer Academy (NCOA) at Flint Kaserne, Bad Toelz, Germany (Figure 6), established under Brigadier General Bruce C. Clarke. Emphasis on NCO education increased to the point that by 1959, over 180,000 Soldiers would attend NCO academies located in the continental United States. In addition to NCO academies, the Army encouraged enlisted men to advance their education by other means.[10] By 1952, the Army had developed the Army Education Program to allow Soldiers to attain credits for academic training. This program provided a number of ways for the enlisted man to attain a high school or college diploma.

Figure 6. Seventh Army NCO Academy

i. In 1958, the Army added two grades to the NCO ranks. These pay grades, E-8 and E-9, would "provide for a better delineation of responsibilities in the enlisted structure."[11] With

---

[10] L.R. Arms, *A Short History of the NCO*, U.S. Army Sergeants Major Academy, 1 November 1991.
[11] Ibid

the addition of these grades, the ranks of the NCO were Corporal (CPL), Sergeant (SGT), Staff Sergeant (SSG), Sergeant First Class (SFC), Master Sergeant (MSG), and Sergeant Major (SGM).

j. In 1966, General Ralph E. Haines Jr., Commanding General, 1966 Continental Army, set-up the NCO Education System (NCOES) and was the driving force in establishing the United States Army Sergeants Major Academy, Fort Bliss, Texas, which conducted its first class in January 1973. In 1987, the Army completed work on a new state-of-the-art training facility at the Sergeants Major Academy at Fort Bliss, Texas, further emphasizing the importance of professional education for NCOs. This 17.5 million-dollar, 125,000 square foot structure allowed the academy to expand course loads and number of courses. As the NCOES continues to grow, the NCO of today combines history and tradition with skill and ability to prepare for combat. She/he retains the duties and responsibilities given to her/him by Von Steuben in 1778, and these have been built upon to produce the Soldier of today.

k. **The Creed of the Noncommissioned Officer.** Written in 1973, the creed (Figure 7) provides SGTs with inspiration to lead while creating a *"yardstick by which to measure themselves."* [12] The creed implies responsibilities for conducting NCO business.

Creed of the Noncommissioned Officer

No one is more professional than I. I am a Noncommissioned Officer, a leader of Soldiers. As a Noncommissioned Officer, I realize that I am a member of a time honored corps, which is known as "the Backbone of the Army." I am proud of the Corps of Noncommissioned Officers and will at all times conduct myself so as to bring credit upon the Corps, the military service and my country regardless of the situation in which I find myself. I will not use my grade or position to attain pleasure, profit or personal safety.

Competence is my watch-word. My two basic responsibilities will always be uppermost in my mind – accomplishment of my mission and the welfare of my Soldiers. I will strive to remain technically and tactically proficient. I am aware of my role as a Noncommissioned Officer. I will fulfill my responsibilities inherent in that role. All Soldiers are entitled to outstanding leadership; I will provide that leadership. I know my Soldiers and I will always place their needs above my own. I will communicate consistently with my Soldiers and never leave them uninformed. I will be fair and impartial when recommending both rewards and punishment.

Officers of my unit will have maximum time to accomplish their duties; they will not have to accomplish mine. I will earn their respect and confidence as well as that of my Soldiers. I will be loyal to those with whom I serve; seniors, peers and subordinates alike. I will exercise initiative by taking appropriate action in the absence of orders. I will not compromise my integrity, nor my moral courage. I will not forget, nor will I allow my comrades to forget that we are professionals, Noncommissioned Officers, leaders!

Figure 7. The Creed of the Noncommissioned Officer

(1) **NCO Subcommittee.** Of those working on the challenges at hand, one of the only NCO-pure instructional departments at the U.S. Army Infantry School (USAIS) at Fort

[12] CSM(R) Dan Elder and CSM(R) Felix Sanchez, "The History of the NCO Creed," *NCO Journal*, Summer 1998.

Benning, Georgia, was the NCO Subcommittee of the Command and Leadership Committee in the Leadership Department. Besides training Soldiers at the NCOA, these NCOs also developed instructional material and worked as part of the team developing model leadership programs of instruction. During one brainstorming session, SFC Earle Brigham recalls writing three letters on a plain white sheet of paper... N-C-O. From those three letters, they began to build the Creed of the Noncommissioned Officer. The idea behind developing a creed was to give NCOs a "yardstick by which to measure themselves." When it was ultimately approved, the Creed of the Noncommissioned Officer was printed on the inside cover of the special texts issued to students attending the NCO courses at Fort Benning, beginning in 1974. Though the Creed of the Noncommissioned Officer was submitted higher for approval and distribution Army-wide, it was not formalized by an official Army publication until 11 years later.

(2) On 15 October 1985, Col Kenneth W. Simpson, Chief, Training and Education, Office of the Chief of Staff, NCOs professional development study, recommended changes to the March 1980 Army NCO Guide, which included the Creed of the Noncommissioned Officer.[13] The Creed of the Noncommissioned Officer was published on 13 November 1986.[14] For over 40 years the Creed of the Noncommissioned Officer has remained relevant because its writers successfully identified what a SGT must *"**Be, Know, and Do**"* as defined in the Army Leadership requirement Model. [15] Written as a concise statement of leadership principles rather than a detailed description of roles and responsibilities, the Creed of the Noncommissioned Officer remains as relevant today as it did when first written and will be relevant as we evolve to meet the changing needs of a NCO Corps of the future.

---

[13] L.R. Arms, *A Short History of the NCO*, U.S. Army Sergeants Major Academy, 1 November 1991.

[14] CSM(R) Dan Elder and CSM(R) Felix Sanchez, "The History of the NCO Creed," *NCO Journal*, Summer 1998.

[15] U.S. Department of the Army, *Army Leadership*, ADP 6-22, 1 August 2012.

(3) The Creed of the NCO will stay relevant as we evolve to meet the changing needs of a modern NCO Corps of 2020. The NCO of 2020 will form an NCO Corps, grounded in the heritage, values, and tradition, which embodies the warrior ethos; values perpetual learning; and is capable of leading, training, and motivating Soldiers.

CHAPTER 2

THE ARMY PROFESSION AND ETHICS

HONORABLE SERVICE

TRUST

STEWARDSHIP OF OUR PROFESSION

ESPRIT DE CORPS

MILITARY EXPERTISE

Chapter 2

The Army Profession and Ethic

2.0. Ref.

- ADP 1, The Army
- ADP 6-22, Army Leadership
- ADRP 1, The Army Profession
- ADRP 6-22, Army Leadership
- AR 600-20, Army Command Policy
- AR 600-25, Salutes, Honors, and Visits of Courtesy
- FM 6-22, Army Leadership: Competent, Confident, and Agile
- http://cape.army.mil/
- http://www.tioh.hqda.pentagon.mil/

2.1. The Army Profession.

a. Our Nation trusts the Army to provide for the national defense. Trust starts with the Oath of Enlistment demonstrating strength of character, commitment to defend the principles of freedom and to fight against tyranny. As a Noncommissioned Officer, it is your duty to carry out the missions assigned to you in accordance with the law and intent of Congress. The foundation on which the Army is built is based on trust. As a leader, you assure your leaders and Soldiers of your competence, character and commitment. Trust is intangible, but your ability to fulfill your roles and discharge your responsibilities depends on the trust between and among Soldiers, between Soldiers and leaders, and between Families and the Army[16]. Our Nation trusts the Army to provide for the national defense. Trust started when you took the Oath of Enlistment (Figure 8) that demonstrated your strength of character; commitment to defend the principles of freedom; and to fight against tyranny. Article VI of the Constitution requires that every member of the Army "shall be bound by Oath or Affirmation, to support the Constitution."[17]

---

[16] U.S. Department of the Army, *The Army,* ADP 1, 6 August 2013.

[17] U.S. Department of the Army, *The Army,* ADP 1, 6 August 2013.

b. Taking the oath is a solemn, moral, and sacred commitment between the Nation and your Soldiers and affirms your commitment to devote yourself to selfless service, adherence to orders, and duty. It is your word and bond to comply with the obligations of each stanza. The Oath (Figure 8) is an ethos peculiar to the character, disposition, and values specific to the Army culture. In addition, the Oath is legally binding and mandates that you and your Soldiers are subject to MCM 2012, 10 USC, and FM 27-10.

I, _____, do solemnly swear (or affirm) that I will support and defend the Constitution of the United States against all enemies, foreign and domestic; that I will bear true faith and allegiance to the same; and that I will obey the orders of the President of the United States and the orders of the officers appointed over me, according to regulations and the Uniform Code of Military Justice. So help me God.

Figure 8. Oath of enlistment

c. When you took the initial Oath, you became a member of the Army Profession. As Army Professionals and leaders, all NCOs must **"Stand Strong"** by certifying or recertifying their competence, character, and commitment.[18] Being a professional involves taking advantage of the opportunity, demonstrating the highest degree of honor and assuming responsibility. We are stronger when we develop and maintain professional knowledge; apply combat power according to law and how personnel and units operate in garrison or on the battlefield.

---

[18] Center for the Army Profession and Ethic, "Stand Strong: Senior Leader Guide", 11 December 2013

2.2. The Army Ethic.

a. The Army Ethic is an evolving set of laws, values, and beliefs, deeply embedded within the core of the Army culture and practiced by all members of the Army Profession to motivate and guide the appropriate conduct of individual members bound together by a common moral purpose.[19]

b. Our ethics and values are continuously challenged when engaged in warfare. The heaviest burden of ethical behavior and enforcement rests with small-unit leaders, who maintain discipline and ensure Soldier conduct remains within ethical and moral boundaries. There are five compelling reasons for this. It is important that you and your Soldiers understand and espouse these ideas, which have the most severe impact on our ability to win the hearts and minds of our enemies and to safeguard honorable service.[20]

- Humane treatment of detainees.
- Humane treatment of noncombatants.
- Make ethical decisions in action fraught with consequences.
- Leaders must not tacitly accept misconduct or encourage it.
- Soldiers must live with the consequences of their conduct.

c. **The US Army as a Military Profession.** The Army Profession is a unique vocation of experts certified in the ethical design, generation, support, and application of land power, serving under civilian authority and entrusted to defend the Constitution and the rights and interests of the American. An Army Professional is a Soldier or Army Civilian who satisfies the requirements for certification in Competence, Character, and Commitment.[21]

---

[19] U.S. Department of the Army, *The Army Profession*, ADRP 1, 14 June 2013

[20] U.S. Department of the Army, *The Army,* ADP 1, 6 August 2013.

[21] U.S. Department of the Army, *The Army Profession*, ADRP 1, 14 June 2013

d. This quest is a duty consistent with our shared identity. Articulating and living by the Army Ethic (Figure 9):"

- Inspires and strengthens our shared *identity* as Trustworthy Army Professionals.
- Expresses **Honorable Service** as our *ethical*, *effective*, and *efficient* conduct of the *mission*, performance of *Duty*, and way of life.
- Motivates our *Duty* to continuously develop **Military Expertise** throughout the Army Profession.
- Emphasizes **Stewardship** of our people and resources and enhances **Esprit de Corps**.
- Drives *Character Development* for the Army.
- Reinforces **Trust** within the profession and with the American people.
- Is essential to Mission Command.[22]

Figure 9. Five essential characteristics of the Army profession

"Being an [Army Professional] means a total embodiment of the Warrior Ethos and the Army Ethics. Our Soldiers need uncompromising and unwavering leaders. We cannot expect our Soldiers to live by an ethic when their leaders and mentors are

---

[22] Center for the Army Profession and Ethic, *The Army Ethic White Paper*, 11 July 2014.

not upholding the standard. These values form the framework of our profession and are nonnegotiable."[23]

- SMA Raymond F. Chandler, III
Former Sergeant Major of the Army

2.3. Army Values

a. Army Values coupled with ethics are the foundation of our Profession. Critical to each Soldier's development is learning about and living by Army Values, when in or out of uniform. Self-discovery, determining the character, applying and living the Seven Core Values reinforces trust in our Soldiers.[24]

- *Loyalty*-Bear true faith and allegiance
- *Duty*-Fulfill your duties
- *Respect*-Treat people as they should be treated
- *Selfless Service*-Put the Welfare of the Nation, the Army and your subordinates before your own.
- *Honor*-Live up to Army Values
- *Integrity*-Do what's right, legally and morally
- *Personal Courage*-Face fear, danger or adversity (physical and moral).

b. The following website provides additional information on Army Values. Website: http://www.army.mil/values

2.4. Army Customs, Courtesies and Traditions.

a. What often sets the Army apart as an institution steeped in history is the commitment to observing Army customs and traditions. It is customs and traditions, strange to the civilian eye, but solemn to the Soldier, which keeps the person in uniform motivated during times of peace. In war, they keep

---

[23] SMA Raymond F. Chandler III, "The Profession of Arms and the Professional Noncommissioned Officer," *Military Review- The Profession of Arms Special Edition* (Sep 2011)

[24] U.S. Department of the Army, *Army Leadership*, ADRP 6-22, 1 August 2012.

the warrior fighting at the front. Educating Soldiers on the importance of observing customs and traditions is key to leader development.

b. Customs. Army customs have been handed down over the centuries and add to the interest, pleasure and graciousness of Army life.

c. Many customs compliment military courtesy. The breach of Army customs may bring disciplinary action. The customs of the Army are its common law. Examples of Army customs are:

- Always render a salute if the situation warrants
- Render proper respects to the flag, reveille, and retreat at all times
- Never criticize Leaders, Soldiers, or the Army in public
- Always make proper use of your chain of command
- Make no excuses while taking responsibility for your actions
- Always speak with your own voice

d. Courtesies. Courtesy among members of the Armed Forces is vital to maintaining discipline. Courteous behavior provides a basis for developing good human relations. Mutual courtesy between subordinates and superiors shows the respect to each member of our profession. Military discipline is founded upon self-discipline, respect for properly constituted authority, and embracing the Army Ethic with its supporting individual values. Military discipline will be developed by individual and group training, which is enforced by NCOs to create a mental attitude resulting in proper conduct and prompt obedience to lawful military authority.[25] Some simple examples and visible signs of respect and self-discipline are:

- When talking to an officer of superior rank, stand at attention until directed otherwise.

---

[25] U.S. Department of the Army, *Army Command Policy*, AR 600-20, 6 November 2014.

- When you are dismissed, or when the officer departs, come to attention and salute.
- When speaking to or being addressed by an NCO of superior rank, stand at parade rest until directed otherwise
- When a NCO of superior rank enters the room, the first Soldier to recognize the NCO calls the room to "At ease."[26]
- When an officer of superior rank enters a room, the first Soldier to recognize the officer calls personnel in the room to attention, but does not salute. A salute indoors is rendered only when reporting.
- Walk on the left of an officer or NCO of superior rank.
- When entering or exiting a vehicle, the junior ranking Soldier is the first to enter, and the senior in rank is the first to exit.
- When outdoors and approached by a NCO, you greet the NCO by saying, "Good morning, Sergeant," for example.
- The first person who sees an officer enter a dining facility gives the order "At ease," unless a more senior officer is already present. Many units extend this courtesy to senior NCOs, also.
- When you hear the command "At ease" in a dining facility, remain seated, silent and continue eating unless directed otherwise. [27]

e. Traditions. Tradition is a customary pattern of thought, action and behavior held by an identifiable group of people. It is information, beliefs, and customs handed down by word of mouth or by example from one generation to another without written instruction. Our military traditions are really the "Army Way" of doing and thinking.

(1) Army traditions are the things that everyone in the Army does.

---

[26] U.S. Department of the Army, *Drill and Ceremonies*, TC 3–21.5, 20 January 2012.

[27] Ibid.

(2) Unit traditions are the unique things that units do that other units may or may not do. Some unit traditions are:

- Ceremonial duties. Soldiers of the Old Guard, the 3d Infantry, have been Sentinels of the Tomb of the Unknown Soldier since 1948.
- The green berets of the Army's Special Forces
- Airborne units' maroon beret.
- Cavalry units' spurs and Stetson hats.
- Special designations (authorized unit nicknames) such as "The 7th Cavalry Regiment's Garry Owen."
- Distinctive items of clothing worn in your unit such as headgear, belt buckles, and tankers' boots.
- Unit mottos such as "Victory!" or "Send me!"

f. The Army Flag and its Streamers. Until 1956, no flag represented the Army as a whole. The first official flag was unfurled on 14 June 1956 (Flag Day and the Army Birthday) at Independence Hall in Philadelphia, Pennsylvania. The Army flag (Figure 10) is in the national colors of red, white, and blue with a yellow fringe (Colors). It has a white field with the War Office seal in blue in its center. Beneath the seal is a scarlet scroll with the inscription "United States Army" in white letters. Below the scroll the numerals "1775" appears in blue to commemorate the year in which the Army was created with the appointment of General George Washington as Commander in Chief. The historic War Office seal, somewhat modified from its original, is the design feature that gives to the Army flag its greatest distinction. "The central element is a Roman cuirass, a symbol of strength and defense. The United States flag, of a design used in the formative years of the Nation, and the other flag emphasize the role of the Army in the establishment of and the protection of the Nation. The sword, esponton (a type of half-pike formerly used by subordinate officers), musket, bayonet, cannon, cannon balls, mortar and mortar bombs are representative of traditional Army implements of battle. The drum and drumsticks are symbols of public notification of the Army's purpose and intent to serve the Nation and its people. The Phrygian cap (often called the Cap of Liberty) supported on the point of the unsheathed sword and the motto "This We'll Defend" on a scroll held by the rattlesnake, a symbol depicted on some American colonial flags, signify the Army's constant readiness to defend and preserve the United States."[28] These Army implements are symbols of strength, defense, and notification that signify the Army's purpose and intent to serve the nation and its people with the readiness to defend and preserve these United States of America.

Figure 10. The Army Flag

g. The colors used in the flag were selected for their traditional significance. Red, white, and blue are the colors, of course, of the national flag. Furthermore, those colors symbolize in the language of heraldry, the virtues of hardiness and valor (red), purity and innocence (white), and vigilance, perseverance, and justice (blue). Blue is especially significant since it has been the unofficial color of the Army for more than two hundred years. The placement of the two flags shown on the seal, the organizational and the national flags are reversed in violation of heraldic custom. The placing of the

---

[28] Institute of Heraldry, U.S. Army Flag and Streamers-Flag Information, http://www.tioh.hqda.pentagon.mil/

United States flag on the left (from the flag's point of view) rather than on the right reflected the tendency of the leaders of the Revolutionary War period to discard traditional European concepts. The display of both an organizational color and the national flag (Figure 11) was a common practice of the Continental Army during the Revolutionary War. There are 188 approved campaigns streamer affixed to the Army Flag, and two open campaigns to be closed and added to the flag when they are assigned an end date and represent on-going campaigns from Operation Enduring Freedom and Operation Iraqi Freedom.[29]

h. The National Flag of the United States of America often called the American Flag, The Stars and Stripes, Old Glory, or Red, White and Blue.[30]

Figure 11. Spirit of 76

---

[29] Institute of Heraldry, U.S. Army Flag and Streamers-Flag Information, http://www.tioh.hqda.pentagon.mil/

[30] US Congress, *Our Flag,* Senate Document 105-013, 5 November 1997.

(1) On June 14, 1777, the Second Continental Congress passed the Flag Resolution, which stated: "*Resolved,* That the flag of the thirteen United States be thirteen stripes, alternate red and white; that the union be thirteen stars, white in a blue field, representing a new constellation." Flag Day is now observed on June 14 of each year.[31] While scholars still argue about this, tradition holds that the Continental Army at the Middlebrook encampment first hoisted the new flag in June 1777.

(2) The flag is a living breathing entity that represents the service and sacrifice of all those who have served before us, who now serve and to those who will serve in the future. It is important to understand the rules of displaying the American flag and the customs of honoring it as the symbol of this great country. The flag is the symbol of freedom, a freedom that came at a cost. It is our duty to guard and protect it and never let it fall to the ground.[32] It is important for leaders and Soldiers to understand its role and symbolism:

2.5. Pride and Esprit de Corps.

a. As history, customs, courtesy, and traditions have a major impact on our Army Culture and our nation; we must instill a sense of pride that defines our character. Pride is a state or feeling of being proud. It is something done by or belonging to oneself or something that causes a person to be proud. Pride can have a negative as well as a positive connotation. Having an inflated sense of one's personal status or accomplishments is a negative connotation. Having a sense of attachment towards one's own or another's choice and action, or towards a whole group of people is having a fulfilled feeling of belonging.[33]

b. Some senses of pride are derived from having:

---

[31] US Congress, *Our Flag,* Senate Document 105-013, 5 November 1997.

[32] Institute of Heraldry, U.S. Army Flag and Streamers-Flag Information, http://www.tioh.hqda.pentagon.mil/

[33] Dr. Alison Adams, "Self-Esteem" *Ezine,* April 2012, Issue 28.

(1) Self Pride- Having self-esteem, self-worth, self-respect, and personal value. It is an essential human need that is vital for survival and normal, healthy development. It arises automatically from within based on your personal beliefs and consciousness. It occurs in conjunction with your thoughts, behaviors, feelings, and actions.

(2) Civic Pride- Is being proud of or relating to a city or town or the people who live therein, or being involved in community affairs. It relates to citizenship and being a citizen. Voting is an example of a civic duty that gives you a sense of civic pride. Taking pride in your unit, your organization, or installation is demonstrating a sense of civic pride.

(3) National Pride- Being proud of your country or proud of yourself as seen in Figure 12. It means being proud and happy to be a citizen. It can take the form of defending your country in times of need and standing by your country even in difficult times.

Figure 12. Honor Guard-Tomb of the Unknown Soldier

(4) Esprit de Corps - Soldiers want to know they are part of a long-standing tradition. Customs and traditions remind them they are the latest addition to a long line of Soldiers. The sense of belonging lives in many veterans long after they have left service. Instilling a sense of esprit comes· by making Soldiers realize they are part of something greater

than themselves. They build deeper Army Values, personal values, family bonds, stronger work ethic and high integrity. It is, therefore, important for leaders to pass on the history that surrounds the organization's crests, awards, decorations, and badges. Upholding traditions ensures the Army's culture becomes indispensable to every member of the Army team.[34]

2.6. Drill and Ceremonies.

a. While we no longer use drill and formations to align the ranks as was done for the phalanxes of Rome or the squares of Waterloo, drill and ceremony is still the foundation of instilling and developing discipline in any size unit and the individual. Additionally, it is still one of the finest methods for developing confidence and troop leading abilities in our subordinate leaders.

b. Drill enables commanders to quickly move their forces from one point to another, mass their forces into a battle formation that affords maximum firepower, and maneuver those forces as the situation develops. The hallmark of the world's best fighting organizations--the Roman legions, the Spartans, the Foreign Legion, the British Brigade of Guards, and many others--is that they are as good on parade as they are in the field or in the attack. The objectives accomplished by drill--professionalism, teamwork, confidence, pride, alertness, attention to detail, esprit de corps, and discipline--are just as important to the modern Army as they were to the militaries of the past.

(1) Drill was historically used to prepare troops for battle.

(2) The three methods of instruction used to teach drill to Soldiers are Step-by-step, Talk-through, and by the numbers. [35]

---

[34] U.S. Department of the Army, *Army Leadership*, ADRP 6-22, 1 August 2012.

[35] U.S. Department of the Army, *Drill and Ceremonies,* TC 3–21.5, 20 January 2012.

(3) A drill command is an oral order of a commander or leader. The precision with which a movement is executed is affected by the manner in which the command is given. [36]

c. Military ceremonies serve several purposes. The following is not an exhaustive list. Military ceremonies can honor high-ranking commanders, officials, or dignitaries; or permit them to observe the state of training of an organization. Military ceremonies can also be used to present decorations and awards, honor or recognize unit or individual achievements, commemorate events, mark changes of command and responsibility, and induct newly promoted NCOs into the ranks.

(1) Military music was used for signaling during encampments, parades and combat.

(2) Bugle Calls were adopted during the Continental Army's contact with the Soldiers and Armies of Europe.

(3) Our National Anthem officially became the "Star Spangled Banner" by law on 3 March 1931.

(4) The Army Song: "The Army Goes Rolling Along" was formally dedicated by the Secretary of the Army on Veterans Day, 11 November 1956.

d. The unit guidon and organizational colors remain an integral part of ceremonies. The art of executing drill with units in unison demonstrate the level of professionalism, discipline and sense of pride Soldiers have in their organizations. Training all Soldiers on the Manual for Guidon is an invaluable skill that instills that sense of pride and professionalism in them.

e. **NCO Induction Ceremony**. The NCO induction ceremony is meant to celebrate the transition of a Soldier to a leader as they join the ranks of a professional NCO corps; the induction ceremony should in no way be used as an opportunity for hazing, but more as a rite of passage. It allows

---

[36] Ibid.

fellow NCOs of a unit to build and develop a cohesive bond and support team development. The importance of recognizing the transition from Soldier to NCO should be shared among superiors, peers and Soldiers of the newly promoted. The induction ceremony should be held separately and serve as an extension of the promotion ceremony. An example of an NCO induction ceremony can be found at the following website:**https://usasma.bliss.army.mil/downloads/nco_induction_ceremony.doc** or **https://atn.army.mil/dsp_template.aspx?dpID=423**

f. **Change of Responsibility.** This ceremony as seen in figure Figure 13 marks the transfer of responsibility for the accomplishments of the unit and the welfare of its Soldiers and their Families from the outgoing to the incoming senior NCO of an organization. An example of a Change of Responsibility Ceremony invitation, program, layout diagram, operations order, request for support, and sword ceremony templates can be found at the following website: **https://atn.army.mil/media/dat/NCO-Corner(HTML)/cocr.aspx**

Figure 13. Change of Responsibility

Note: for more information on retirement and memorial ceremonies please visit the following link https://armypubs.us.army.mil/doctrine/DR_pubs/dr_aa/pdf/tc3_21x5.pdf

2.7. Inspections.

a. Military inspections were created by Peter I of Russia for checking on conditions of unit administration, services and material supply of the troops.[37] Today, they provide the means to assess the capabilities of an organization and to identify any potential problems. NCOs are vital to the inspection program to ensure the unit is operating efficiently, effectively, and free of issues detrimental to readiness, morale, and mission accomplishment.

b. Inspections must have a specific purpose.

(1) Be related to mission accomplishment.
(2) Tailored to meet the commander's needs.

(3) Be performance oriented and start with an evaluation against recognized standards in order to identify compliance with that standard.

(4) Capable of identifying and analyzing process improvement opportunities that will increase performance support, transformation, and reduce risks (Inspections, 2014). Basic elements of an inspection are:

(a) Measure performance against a standard.

(b) Determine the magnitude of the problem(s).

(c) Seek the root cause(s) of the problem(s).

(d) Determine a solution.

(e) Assign responsibility to the appropriate individuals or agencies.

c. **Organizational Inspection Program (OIP).** The OIP provides the commander with an organized management tool to identify, prevent or eliminate problem areas.

---

[37] David R. Stone, *A Military History of Russia*, 2006.

d. **Command Inspections.**

(1) Command Inspections ensures units comply with regulations and policies and allow commanders to hold leaders at all levels accountable for this compliance. It allows commanders to determine the training, discipline, readiness, and welfare of the command and help identify systemic problems within the units and assist in the recognition of emerging trends (Army Inspection, 2014).

(2) Staff Assistance Visits (SAV). SAVs provide assistance in unit functional areas such as maintenance, logistics, human resources, and training to improve and refine processes and to prepare for inspections of those functional areas. SAV are intended to help the unit help itself.

2.8. What makes "YOU" an Army Professional? NCOs are certified in Competence, Character, and Commitment. Consistent demonstration of these qualities develops mutual trust within cohesive teams.

- **Competence.** Demonstrated ability to successfully perform duty with discipline and to standard.
- **Character.** Dedication and adherence to the Army Ethic, including Army Values, as consistently and faithfully demonstrated in decisions and actions.
- **Commitment.** Resolve to contribute Honorable Service to the Nation and accomplish the mission despite adversity, obstacles, and challenges.

CHAPTER 3

MISSION COMMAND

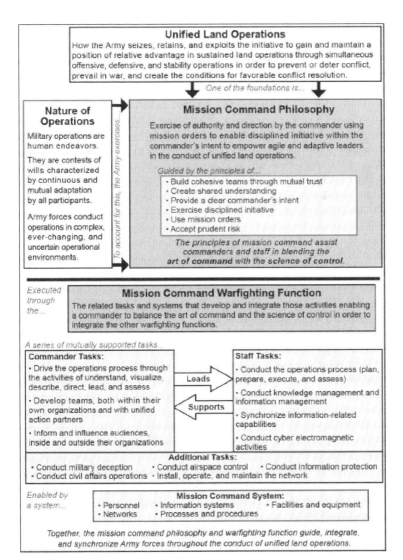

Unified Land Operations
How the Army seizes, retains, and exploits the initiative to gain and maintain a position of relative advantage in sustained land operations through simultaneous offensive, defensive, and stability operations in order to prevent or deter conflict, prevail in war, and create the conditions for favorable conflict resolution.

One of the foundations is...

Nature of Operations
Military operations are human endeavors.

They are contests of wills characterized by continuous and mutual adaptation by all participants.

Army forces conduct operations in complex, ever-changing, and uncertain operational environments.

To account for this, the Army exercises...

Mission Command Philosophy
Exercise of authority and direction by the commander using mission orders to enable disciplined initiative within the commander's intent to empower agile and adaptive leaders in the conduct of unified land operations.

Guided by the principles of...
- Build cohesive teams through mutual trust
- Create shared understanding
- Provide a clear commander's intent
- Exercise disciplined initiative
- Use mission orders
- Accept prudent risk

The principles of mission command assist commanders and staff in blending the art of command with the science of control.

Executed through the...

Mission Command Warfighting Function
The related tasks and systems that develop and integrate those activities enabling a commander to balance the art of command and the science of control in order to integrate the other warfighting functions.

A series of mutually supported tasks...

Commander Tasks:
- Drive the operations process through the activities of understand, visualize, describe, direct, lead, and assess
- Develop teams, both within their own organizations and with unified action partners
- Inform and influence audiences, inside and outside their organizations

Leads

Supports

Staff Tasks:
- Conduct the operations process (plan, prepare, execute, and assess)
- Conduct knowledge management and information management
- Synchronize information-related capabilities
- Conduct cyber electromagnetic activities

Additional Tasks:
- Conduct military deception
- Conduct civil affairs operations
- Conduct airspace control
- Install, operate, and maintain the network
- Conduct information protection

Enabled by a system...

Mission Command System:
- Personnel
- Networks
- Information systems
- Processes and procedures
- Facilities and equipment

Together, the mission command philosophy and warfighting function guide, integrate, and synchronize Army forces throughout the conduct of unified land operations.

Chapter 3.

Mission Command (MC).

3.0. Ref.

- ADP 3-0, Unified Land Operations
- ADP 6-0, Mission Command
- ADRP 3-0, Unified Land Operations
- ADRP 6-0, Mission Command
- FM 6-0, Commander and Staff Organization and Operations
- **https://atn.army.mil/media/dat/MCResources/mcr esources.aspx?**

3.1. The Army's primary mission is to organize, train, and equip forces to conduct prompt and sustained land combat operations. The Army does this through its operational concept of unified land operations.[38]

a. **Unified Land Operations.** ULO (Figure 15) is the concept used to describe how the Army seizes, retains and exploits the initiative to gain and maintain a position of relative advantage in sustained land operations through simultaneous offensive, defensive, and stability operations in order to prevent or deter conflict, prevail in war, and create conditions for favorable conflict resolution.[39]

---

[38] U.S. Department of the Army, *Unified Land Operations*, ADRP 3-0, 16 May 2012.

[39] U.S. Department of the Army, *Unified Land Operations*, ADP 3-0, 10 October 2011.

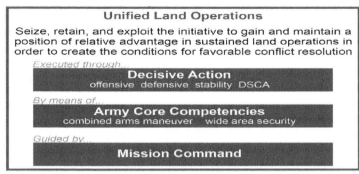

Figure 15. Unified Land Operations

b. **Military operations are human endeavors.** They are contests of wills characterized by continuous and mutual adaptation by all participants. Army forces conduct operations in complex, ever-changing, and uncertain operational environments (Figure 16).[40] During operations, unexpected opportunities and threats rapidly present themselves. Operations require responsibility and decision-making at the point of action. Commanders seek to counter the uncertainty of operations by empowering subordinates at the scene to make decisions, act, and quickly adapt to changing circumstances. Through mission command, commanders initiate and integrate all military functions and actions toward a common goal – mission accomplishment.

Nature of Operations

Military operations are **human endeavors.** They are contests of wills characterized by continuous and mutual adaptation by all participants. Army forces conduct operations in **complex, ever-changing, and uncertain** operational environment.

To cope with this, the Army exercises ...

c. **Figure 16. Nature of Operations**

---

[40] U.S. Department of the Army, *Mission Command,* ADRP 6-0, 17 May 2012.

3.2. Mission command philosophy. Mission command (Figure 17) is the exercise of authority and direction by the commander. It requires using mission orders to enable disciplined initiative within the commander's intent to empower agile and adaptive leaders in the conduct of unified land operations. The mission command philosophy effectively accounts for the nature of military operations. Throughout operations, unexpected opportunities and threats rapidly present themselves. Operations require responsibility and decision making at the point of action. Through mission command, commanders initiate and integrate all military functions and actions toward a common goal—mission accomplishment.[41]

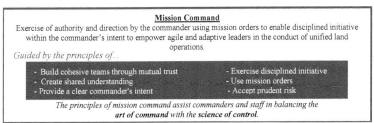

Mission Command

Exercise of authority and direction by the commander using mission orders to enable disciplined initiative within the commander's intent to empower agile and adaptive leaders in the conduct of unified land operations.

Guided by the principles of...

- Build cohesive teams through mutual trust	- Exercise disciplined initiative
- Create shared understanding	- Use mission orders
- Provide a clear commander's intent	- Accept prudent risk

The principles of mission command assist commanders and staff in balancing the **art of command** *with the* **science of control.**

Figure 17. Mission Command

3.3. The Six Principles of Mission Command.

- **Build cohesive teams through mutual trust:** by upholding Army Values and exercising leadership consistent in the Army's leadership principles.
- **Create shared understanding and purpose:** by maintaining collaboration and dialogue throughout the operations process
- **Provide a clear commander's intent, purpose, key tasks, desired end state and resources.** By giving the reason for the operation, conveys a clear image of the operation's purpose, key tasks, and desired outcome.
- **Exercise disciplined initiative by taking action to develop the situation.** Disciplined initiative is action

---

[41] U.S. Department of the Army, *Mission Command,* ADP 6-0, 17 May 14.

in the absence of orders, when existing orders no longer fit the situation, or when unforeseen opportunities or threats arise.

- **Use Mission Orders:** directives that emphasize the results to be attained.
- **Accept prudent risk:** is the deliberate exposure to potential injury or loss when the terms of mission accomplishment are worth the cost.

3.4. The Art of Command. Command is the authority that a commander lawfully exercises over subordinates by virtue of rank or assignment. The Art of Command is the creative and skillful exercise of authority through timely decision-making and leadership. As an art, command requires the use of judgment.

3.5. The Science of Control. Control is the regulation of forces and war fighting functions to accomplish the mission in accordance with the commander's intent. Control permits commanders to adjust operations to account for changing circumstances and direct the changes necessary to address the new situation.

3.6. Mission Command Warfighting Function.

a. Warfighting function (Figure 18) is the related tasks and systems that develop and integrate those activities enabling a commander to balance the art of command and the science of control in order to integrate the other warfighting functions.[42] As a warfighting function, mission command consists of the related tasks and a mission command system that support the exercise of authority and direction by the commander. Through the mission command warfighting function, commanders integrate the other warfighting functions into a coherent whole to mass the effects of combat power at the decisive place and time.

---

[42] U.S. Department of the Army, *Mission Command*, ADRP 6-0, 17 May 2012.

Figure 18. Mission Command Warfighting Function

b. The commander tasks are:

- Drive the operations process through activities of understanding, visualizing, describing, directing, leading, and assessing operations.
- Develop teams, both within their own organizations and with unified action partners.
- Inform and influence audiences, inside and outside their organization.

c. The staff tasks are:

- Conduct the operations process: plan, prepare, execute and assess (Figure 19).
- Conduct knowledge management and information management.
- Synchronize information-related capabilities.
- Conduct cyber electromagnetic activities.

d. The Army must shape and train for 21st Century Operations. MC is essential to shaping the Army, driving change and adapting the Army, which unites efforts to effectively synchronize and integrate operational and generating forces' roles and responsibilities as applied against Doctrine, Organization, Training, Materiel, Leadership and Education, Personnel, Facilities and Policy (DOTMLPF-P).

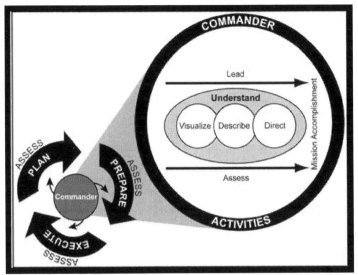

Figure 19. The Operations Process

3.7. NCO responsibilities to Mission Command

- Exercise disciplined initiative to respond to unanticipated problems.
- Be prepared to assume responsibility.
- Maintain unity of effort.
- Take prudent action.
- Understand the Commander's intent.
- Understand the overall common objective.
- Act resourcefully within the commander's intent.
- Act and synchronize actions with the rest of the force.
- Enforce standards and discipline while conducting daily missions and making decisions.
- Live by and uphold the Army Ethic, be a standard-bearer and lead by example.
- Care for Soldiers and set the example for them.
- Develop subordinates.
- Advise commanders at all levels
- Support the full range of Joint Force Commanders' future requirements, creating opportunities to better achieve national objectives

This page intentionally left blank.

Genuine Leadership, the Little Things Matter

Authority

Army Values

Custom and Courtesies

Emotional Intelligence and Leadership

Engaged Leadership

Problem Solving

From one Leader
to Another

Chapter 4

The Roadmap to Leadership

4.0. Ref.

- ADP 1, The Army
- ADP 6-22, Army Leadership
- ADP 7-0, Training Units and Developing Leaders
- ADRP 1, The Army Profession
- ADRP 6-22, Army Leadership
- ADRP 7-0, Training Units and Developing Leaders
- AR 27-10, Military Justice
- AR 350-1, Army Training and Leader Development
- AR 600-9, The Army Body Composition Program
- AR 600-20, Army Command Policy
- AR 670-1, Wear and Appearance of Army Uniforms and Insignia
- ATP 6-22.1, The Counseling Process
- DA Pam 611-21, Military Occupational Classification and Structure
- FM 6-22, Army Leadership; Competent, Confident, and Agile

4.1. This chapter is intended to provide a roadmap to leading (Figure 20) and provides you with the tools to make you aware of who you are, how others view you, and how you view others. It is important that you know yourself and know your Soldiers. Once you understand who you are, you will be better prepared to develop and lead your Soldiers. It will give you an understanding of how to identify your leadership capabilities and to improve on them. Additionally, this chapter outlines content and resources for your self-development. The readings and references provide more comprehensive details on the subjects that best establish the foundation of the art and science of being a leader. This guide is not all-inclusive, but provides you a pathway to doctrine, policy, and procedures that will make you a more effective leader.

Figure 20. Roadmap to Effective Leadership

4.2. To be a competent and confident leader, you must know the scope of your duties and responsibilities, the meaning of those duties, and the benefits or consequences of your actions in exercising your responsibilities. As a leader who trains, cares, and develops Soldiers, you will take on the role of trainer, disciplinarian, coach, and leader. Your Soldiers will trust you when they can rely on your Competence, Character, and Commitment to live by and uphold the Army Ethic.

4.3. Army Leadership Requirements. Leadership is the process of influencing people by providing purpose, direction, and motivation to accomplish the mission and improve the organization. Accomplishing the mission however is not enough; the leader is responsible for developing individuals and improving the organization for the near- and long-term.[43] The role of NCO leader is to:

- Convey information and provide day-to-day guidance.
- Train Soldiers to cope, prepare, and perform regardless of the situation.

---

[43] U.S. Department of the Army, *Army Leadership*, ADP 6-22, 1 August 2012.

- Set and maintain high-quality standards and discipline.
- Be a standard-bearer and role model critical to training, educating, and developing subordinates.
- Prepare Soldiers for missions, stressing tradecrafts and physical hardening.
- Coach, counsel and mentor.

4.4. "Effective leadership blends core leader competencies (groups of related leader behaviors that lead to successful performance and are common throughout the organization and consistent with the organization's mission and value) with leadership attributes (characteristics inherent to the leader that moderates how well learning and performance occur)."[44] The Army leader's attributes and competencies are listed in Figure 21.[45]

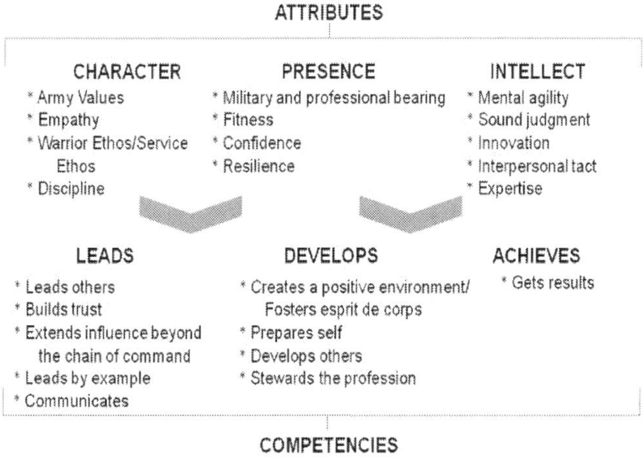

ATTRIBUTES

CHARACTER	PRESENCE	INTELLECT
* Army Values	* Military and professional bearing	* Mental agility
* Empathy	* Fitness	* Sound judgment
* Warrior Ethos/Service Ethos	* Confidence	* Innovation
* Discipline	* Resilience	* Interpersonal tact
		* Expertise

LEADS	DEVELOPS	ACHIEVES
* Leads others	* Creates a positive environment/ Fosters esprit de corps	* Gets results
* Builds trust	* Prepares self	
* Extends influence beyond the chain of command	* Develops others	
* Leads by example	* Stewards the profession	
* Communicates		

COMPETENCIES

Figure 21. The Army Leadership Requirements Model

4.5. Attributes of Leading. There are three categories of attributes that affect the actions that leaders perform and shape how they behave and learn in their environment. These attributes capture a leader's values, demeanor, and their

---

[44] Center for Army Leadership, *Leader Development Improvement Guide*, November 2014.

[45] Ibid.

mental and social faculties applied in the act of leading.[46] Attributes are key ingredients in accomplishing your goals. Here is where you as a leader must provide a clear and consistent way of conveying your expectations so that those you lead understand your intent. Attributes affect the actions that leaders perform. Good character, solid presence and keen intellect enable the core leader competencies to be performed with greater effect. The three categories of attributes are:

- **Character:** Army Values, Empathy, Warrior Ethos, Service Ethos, and Discipline
- **Presence:** Military and professional bearing, fitness, confidence, and resilience
- **Intellect:** mental agility, sound judgment, innovation, interpersonal tact, and expertise

4.6. Competencies. There are three core categories of competencies according to Army Leadership, ADP and ADRP 6-22. The Army leader serves to lead others, to develop the environment, themselves, others and the profession as a whole, and to achieve organizational goals. Competencies provide a clear and consistent way of conveying expectations for Army leaders. There are a total of 10 individual competencies that comprise the three main core categories. **http://armypubs.army.mil/doctrine/DR_pubs/dr_a/pdf/adp 6_22_new.pdf**

    a. Lead: The ***lead category*** encompasses five competencies: leads others, extends influence beyond the chain of command, builds trust, leads by example, and communicates. Army leaders apply character, presence, and intellect to the core leader competencies while guiding others toward a common goal and mission accomplishment. Direct leaders influence others through person-to-person techniques and/or actions, such as a team leader who instructs, encourages hard work, and recognizes achievement. Organizational and strategic leaders guide their organizations using indirect means of influence. At every level, leaders take advantage of formal and informal processes to extend

---

[46] U.S. Department of the Army, *Army Leadership*, ADP 6-22, 1 August 2012.

influence beyond the traditional chain of command. A detailed description of the *lead* core competency and its sub-competencies can be found in chapter six, ADRP 6-22, Army Leadership.
http://armypubs.army.mil/doctrine/DR_pubs/dr_a/pdf/adrp6_22_new.pdf

    b. Develop: The ***develop category*** encompasses four competencies: <u>create a positive environment</u>, <u>prepares self</u>, <u>develop others</u>, and <u>stewards the profession</u>. Effective leaders strive to leave an organization better than they found it and expect other leaders to do the same. Leaders have the responsibility to create a positive organizational climate, prepare themselves to do well in their duties, and help others to assume positions with greater leadership responsibility. They work on self-development to prepare for new challenges. A detailed description of the *develop* core competency and its sub-competencies can be found in chapter seven, ADRP 6-22, Army Leadership.
http://armypubs.army.mil/doctrine/DR_pubs/dr_a/pdf/adrp6_22_new.pdf

    c. Achieve: The ***achieve category*** contains the single competency of <u>gets results</u>. Leadership builds effective organizations. Effectiveness directly relates to the core leader competency of getting results. From the definition of leadership comes achieving focuses on accomplishing the mission. Mission accomplishment co-exists with an extended perspective towards maintaining and building the organization's capabilities. Achieving begins in the short-term by setting objectives. In the long-term, achieving requires getting results in pursuit of those objectives. Getting results focuses on structuring what to do to produce consistent results. A detailed description on the *achieve* core competency and its sub-competency can be found in chapter eight ADRP 6-22, Army Leadership.
http://armypubs.army.mil/doctrine/DR_pubs/dr_a/pdf/adrp6_22_new.pdf

    d. NCO leader competencies improve over time. NCOs continuously refine and extend their ability to perform these competencies proficiently and learn to apply them in increasingly complex situations. Leaders do not wait until combat deployments to develop their leader competencies; they are developed, sustained, and improved upon by performing one's assigned tasks and missions daily. They use every peacetime training opportunity to assess and improve their ability to lead Soldiers. NCOs improve proficiency of their competencies by taking advantage of every opportunity to learn and grow. They look for new learning opportunities, ask questions, seek out additional training opportunities, and request performance critiques from supervisors, peers, and subordinates alike. This lifelong approach to learning ensures that our NCOs remain viable as a professional corps. A more detailed explanation of each of the competencies listed can be found in ADP/ADRP 6-22, Army Leadership.

4.7. 21st Century Soldier Competencies.

a. All Soldiers and leaders must master the fundamental warrior skills supporting tactical and technical competence to execute full-spectrum operations among diverse cultures, with joint, interagency, intergovernmental, and multinational partners, at the level appropriate for each cohort and echelon. The learning environment and instructional strategies must simultaneously integrate and reinforce competencies that develop adaptive and resilient Soldiers and leaders of character who can think critically and act ethically. The nine 21st century competencies (Figure 22) are:

- Character and accountability
- Comprehensive fitness
- Adaptability and initiative
- Lifelong learner (includes digital literacy)
- Teamwork and collaboration
- Communication and engagement (oral, written, negotiation)
- Critical thinking and problem solving
- Cultural and joint, interagency, intergovernmental, and multinational competence
- Tactical and technical competence (full spectrum capable)

Figure 22. The NCO 4 x 6 Model

b. An NCO should be:

- Professional Soldier
- Mentor
- Comprehensively Fit
- Ethical Decision Maker
- Competent
- Proficient in Duties
- Self-Aware
- Responsible
- Servant Leader
- Effective Communicator
- Strength of Character
- Loyal
- Critical Thinker/Problem Solver
- Dependable & Accountable
- Moral

4.8. Army Leadership Levels. Figure 23 illustrates the three levels of leadership: direct, organizational, and strategic. Factors shaping a position's leadership level can consist of the position's span of control, its headquarters level, degree of control used by the leader holding the position, size of the unit or organization, type of operations, and its planning horizon.[47]

Figure 23. Army Leadership Levels

a. **Direct Leadership.** "Direct leadership is face-to-face or first-line leadership. It generally occurs in organizations where subordinates see their leaders all the time: teams, squads, sections, platoons, departments, companies, batteries, and troops. The direct leader's span of influence may range from a few to dozens of people".[48]

b. **Organizational Leadership.** "Organizational leaders influence several hundred to several thousand people. They do this indirectly, generally through more levels of subordinates and staffs than do direct leaders. The additional levels of subordinates can make it more difficult for them to see and judge immediate results."[49]

c. **Strategic Leadership.** The "strategic leaders include military and civilian leaders at the major command through DOD levels. Strategic leaders are responsible for large organizations and influence several thousand to hundreds of

---

[47] U.S. Department of the Army, *Army Leadership*, ADRP 6-22, 10 September 2012.

[48] Ibid.

[49] Ibid.

thousands of people. They establish force structure, allocate resources, communicate strategic vision, and prepare their commands and the Army for future roles" [50]

4.9. Self-Awareness/Self-Discovery

a. One's self-awareness is about preparing self. Self-awareness enables leaders to recognize their strengths and weaknesses. As a leader, you must be able to form accurate self-perceptions, other's perceptions of you and change as appropriate.[51] Self-awareness is about developing a clear, honest picture of your capabilities and limitations. Leaders of character who embrace Army leader attributes and competencies will be authentic, positive leaders.

b. To be an effective leader, you must know yourself, your strengths, weaknesses, and capabilities in order to better understand your personality. Personality is a set of character traits and tendencies that determine common or different psychological behaviors such as your feelings, thoughts, and competencies. Consider the following questions when reflecting on who you are:

- Who you are in any given situation?
- How do you behave in any given situation?
- What habits do you possess?
- What choices and decisions do you make?
- How do you interact with others and what is your role in your social environment?
- What tendencies were you born with and how have those tendencies matured?
- What influenced you as you adapted, grew, and developed?
- What did you learn that shaped your view of the world?
- What is your view of the world?
- What are your traits?

---

[50] U.S. Department of the Army, *Army Leadership*, ADRP 6-22, 1 August 2012.

[51] Colonel David A. Lesperance, *Developing Leaders for Army 2020*, U.S. Army War College, Class 2012

- How do your character traits fit into the Army?
- What TYPE personality are you?

c. We all judge others by their behavior, but we seldom understand the intent or motivation behind those behaviors. Knowing your personality, helps you understand what is behind your outward behavior or how people view you. In order for you to understand yourself, you must self-reflect and self-discover. There are several publications to help you determine your personality type to include the Army Multi-source Assessment and Feedback tools. Refer to the recommended professional reading list at Appendix A for publications to aid you in self-discovery.

4.10. Multi-Source Assessment and Feedback (MSAF) and 360 assessment. MSAF is an assessment tool that promotes self-awareness for individual leader development. MSAF provides feedback to leaders from their subordinates, peers and leaders as they relate to the eight leadership competencies in ADRP and FM 6-22. The objective is to purge the ranks of toxic leaders. It is to promote self-awareness and individual development that leads to skill improvement, adaptability, and better performance. Every NCO is required to take the MSAF prior to attending NCOES courses. More information and to enroll into MSAF visit the MSAF website at **https://msaf.army.mil**

> **"I believe that multi-dimensional feedback is an important component to holistic leader development. By encouraging input from peers, subordinates and superiors alike, leaders can better "see themselves" and increase self-awareness. A 360-degree approach applies equally to junior leaders at the squad, platoon, and company level as well as to senior leaders. The ability to receive honest and candid feedback, in an anonymous manner, is a great opportunity to facilitate positive leadership growth."**
>
> **GEN Ray Odierno**
> **Chief of Staff of the Army**

4.11. The Virtual Improvement Center (VIC). The VIC is an online source to training and education products and links to enable you to learn and practice the leadership competencies in ADP 6-22 and ADRP 6-22. The VIC, located at the MSAF website, contains links to improvement materials, applications and resources. Each link represents one of the leader competencies. More information and to access VIC visit the MSAF website at **https://msaf.army.mil**

4.12. Communication. One of the best characteristics of knowing yourself is how well you communicate. *Johari Window* is one of many tools that the Army uses to improve communications. The better prepared you are in regards to communication, the more comfortable you are in opening up to others and become an effective communicator. The more effective you are at listening can improve your effectiveness as a leader. The following website provides additional information to assist you in improving yourself and your communications skills of active listening, starting goals for action and share understanding.
http://www.mindtools.com/CommSkll/CommunicationIntro.htm

4.13. Leader Development. Leader development is a deliberate, continuous, sequential, and progressive process grounded in Army Values. Leader development occurs through lifelong synthesis of education, training and experience. It is about leaders who act with boldness and initiative in dynamic, complex situations to execute missions according to present and future doctrine. The goals of leader development are:

a. Leaders develop subordinates into leaders for the next level.

b. Leaders are inherently Soldiers first who are tactically and technically proficient and adaptive to change.

c. Leaders create competent and confident future leaders, capable of leading training and preparing units for their mission.

d. Leaders must quickly train small-unit leaders to reach proficiency to operate in widely dispersed areas of combined arms teams and integrate with unified action partners.

e. Leaders must be self-aware and adaptive.

f. Leaders must be comfortable with ambiguity, able to anticipate possible second and third order effects.

g. Leaders must be multifunctional to exploit combined arms and joint integration.

4.14. The Army Training and Leader Development Model. Shows the three separate, but overlapping learning domains (operational, institutional, and self-development) used to achieve the goal of trained Soldiers, Army civilians, leaders, and ready units (Figure 24).

Figure 24. The Army Training and Leader Development Model

a. Leader Development is the deliberate, continuous, sequential, and progressive process that develops Soldiers and Army civilians into competent and confident leaders capable of decisive action, mission accomplishment, and taking care of Soldiers and their Families.

b. Leaders gain needed skills, knowledge, and experience through a combination of institutional training education, operational assignments and self-development.[52]

c. An Army leader is anyone who by virtue of assumed role or assigned responsibility inspires and influences people to accomplish organizational goals. Army leaders motivate people both inside and outside the chain of command to pursue actions, focus thinking, and shape decisions for the greater good of the organization.[53] The Leadership Requirements Model (LRM) establishes what attributes and competencies leaders need to become an effective leader. It is achieved through the lifelong synthesis of the knowledge, skills, and experiences gained through the three domains of development.[54]

- Institutional training and education
- Operational assignments
- Self-development.

4.15. The Army Leader: Person of Character, Presence, and Intellect. These three domains interact by using feedback and assessment from various sources and methods, and each training domain complements the other two. All of the domains have an important role in training Soldiers and Army civilians, growing leaders, and preparing units for deployment. It is important to emphasize leaders who:

- Have conviction
- Have character
- Are present
- Exercise intellect

---

[52] U.S. Department of the Army, *Army Training and Leader Development*, AR 350-1, 19 August 2014.

[53] U.S. Department of the Army, *Army Leadership*, ADP 6-22, 1 August 2012.

[54] U.S. Department of the Army, *Army Training and Leader Development*, AR 350-1, 19 August 2014.

- Are servant leaders
- Recommend opinions of others
- Empower subordinates- Step up to lead
- Are open to feedback and seek it
- Adjust their thoughts, feelings and actions called self-regulation
- Show empathy
- Are resilient
- Communicate candidly
- Demonstrate moral courage

a. The Army's formal leader development process promotes the growth of individuals through training and education, experience, assessment, counseling and feedback, remedial and reinforcement actions, evaluation, and selection.

b. Leader Presence. Presence is not just a matter of the leader showing up; it involves the image that the leader projects. A leader's effectiveness is dramatically enhanced by understanding and developing the following areas:

(1) Military bearing: projecting a commanding presence, a professional image of authority. A professional presents a decent appearance because it commands respect.

(2) Physical fitness: having sound health, strength, and endurance, which sustain emotional health and conceptual abilities under prolonged stress. A Soldier is similar to a complex combat system. A Soldier needs exercise, sufficient sleep, and adequate food and water for peak performance.

(3) Confidence: projecting self-confidence and certainty in the unit's ability to succeed in whatever it does; able to demonstrate composure and outward calm through steady control of emotion. Confident leaders help Soldiers control doubt while reducing team anxiety.

(4) Resilience: showing a tendency to recover quickly from setbacks, injuries, adversity, and stress while maintaining a mission and organization focus. Resilience rests on will, the inner drive that compels leaders to keep going, even when exhausted, hungry, afraid, cold, and wet.

4.16. Methods of Influence. The components of leadership involve one who leads, and another who follows and every leader is a follower. You must balance the success of the mission with how you treat and care for your Soldiers. Key elements to applying influences are:

- **Pressure:** when leaders use explicit demands to achieve compliance. Should be infrequently used as it triggers resentment from followers.
- **Legitimating:** when leaders establish their authority as the basis for the request. Certain jobs must be accomplished regardless of circumstances.
- **Exchange:** when leaders make an offer to provide some desired item or action in trade for compliance with a request. Here, the leader controls certain resources or rewards.
- **Personal Appeal:** when the leader asks the follower to comply with a request based on friendship or loyalty. Mutual trust is key to success when faced with difficult situations.
- **Collaboration:** when the leader cooperates in providing assistance or resources to carry out a directive or request. Leaders make the choice more attractive by stepping in and resolving problems.
- **Rational Persuasion:** when the leader provides evidence, logical argument or explanations show how a request is relevant to the goal. This is often the first approach to gaining compliance or commitment from the follower.
- **Apprising:** when the leader explains why the request will benefit the follower. Sometimes the benefits are out of the control of the leader.
- **Inspirational Appeals:** when the leader fires up enthusiasm by arousing strong emotions to build conviction. Here, leaders inspire followers to surpass minimal standards and reach elite performance.
- **Participation:** when the leader asks others to take part in addressing a problem or meet an objective. Involving key leaders at all levels ensures followers take stock in the vision.

4.17. Counseling.

a. Counseling is one of the most important leadership development responsibilities for Army leaders. The Army's future and the legacy of today's Army leaders rest on the shoulders of those they help prepare for greater responsibility. Developing leaders who will replace leaders of today should be one of a leader's highest priorities, and it supports the Warrior Ethos of training and developing Soldiers into warriors. Counseling tells Soldiers what they are doing right, wrong, and how to improve on their weaknesses. This enables them to become disciplined, physically and mentally tough, trained, and proficient in their duties, standing ready to deploy, engage, and destroy the enemy when called to do so.

b. Counseling is central to leader development. It is the process used by leaders to review a subordinate's demonstrated performance and potential. Leaders who serve as designated raters have to prepare their subordinates to be better Soldiers or Army Civilians. Good counseling focuses on the subordinate's performance and issues with an eye toward tomorrow's plans and solutions. Leaders expect subordinates to be active participants seeking constructive feedback. Counseling cannot be an occasional event but should be part of a comprehensive program to develop subordinates. With effective counseling, no evaluation report—positive or negative—should be a surprise. A consistent counseling program includes all subordinates, not just the people thought to have the most potential.

c. One of the key components of the counseling process is educating our NCOs on the process of developing individual development plans and the key components of the developmental counseling process to enhance professional growth and development as well as promotion potential. Additionally, emphasis should be given on the importance and use of the Soldier Leader Risk Assessment tool and how it is integrated into the counseling program.

d. There are three types of developmental counseling.

(1) **Event Counseling.** This involves a specific event or situation. Event oriented counseling includes:

- Instances of superior or substandard performance.
- Reception and Integration counseling.
- Crisis Counseling.
- Referral Counseling.
- Promotion Counseling.
- Separation Counseling.

(2) **Performance Counseling.** When leaders conduct a review of a subordinate's duty performance over a certain period. Performance Counseling includes:

- Discussion of established performance objectives and standards for the next period.
- Periodic performance counseling as part of the NCOER support form requirements.
- Beginning of and during the evaluation period and provides opportunity for leaders to establish and clarify expected values, attributes, and competencies.

(3) **Professional growth counseling.** Professional growth counseling includes:

- Planning for the accomplishment of individual and profession goals.
- Identify and discuss subordinate's strengths and weaknesses.
- Create an individual development plan that builds on those strengths and weaknesses.
- Opportunities for civilian and military schooling, future assignments special programs and reenlistment options.

(4) Refer to ATP 6-22.1 on the qualities of a counselor to include respect for subordinates, self-awareness and cultural awareness, empathy, and credibility. For additional information about a developmental counseling learning lesson, visit the following website: **http://usacac.army.mil/CAC2/cal/dc/cws.htm**

4.18. NCO Support Channel.

a. The NCO Support Channel (leadership chain) parallels and complements the chain of command. It is a channel of communications and supervision from the Command Sergeant Major (CSM) to 1SG and then to other NCOs and enlisted personnel of the unit. Commanders define responsibilities and authority of their NCOs to their staff and subordinates by:

- Transmitting, instilling, and ensuring the efficacy of the professional Army ethic.
- Planning and conducting day-to-day unit operations within prescribed policies and directives.
- Training of enlisted Soldiers in their MOS as well as in the basic skills and attributes of a Soldier.
- Supervising unit physical fitness training and ensuring that unit Soldiers comply with the weight and appearance standards of AR 600–9, and AR 670–1.
- Teaching Soldiers the history of the Army, to include military customs, courtesies, and traditions.
- Caring for individual Soldiers and their Families, both on and off duty.
- Teaching Soldiers the mission of the unit and developing individual training programs to support the mission.
- Accounting for and maintaining individual arms and equipment of enlisted Soldiers and unit equipment under their control.
- Administering and monitoring the Noncommissioned Officer Development Program, and other unit training programs.
- Achieving and maintaining courage, candor, competence, commitment, and compassion.

b. DA Pam 611–21 and AR 600-20 contain specific information concerning the responsibilities, command functions, and scope of NCO roles.

Sergeant Major of the Army

This is the senior SGM grade and designates the senior enlisted position of the Army. The SGM in this position serves as the senior enlisted adviser and consultant to the Secretary of the Army and the Chief of Staff of the Army.

The Command/Sergeant's Major Role

a. The CSM is the senior NCO of the command at battalion or higher levels. The CSM carries out policies and standards on performance, training, appearance and conduct of enlisted personnel. The CSM gives advice and makes recommendations to the commander and staff in matters pertaining to the organization. A unit, installation, or higher headquarters CSM directs the activities of the NCO Support Channel. The support channel functions orally through the CSM or 1SG's call. The CSM administers the unit NCO Development Program (NCODP), normally through written directives and the NCO Support Channel. As the senior NCO of the command, the CSM is the training professional within the unit, overseeing and driving the entire training program. The CSM assists the commander in determining leader tasks and training for NCOs.

b. The SGM is often the key enlisted member of the staff elements at battalion and higher levels. The SGM's experience and ability are equal to that of the unit CSM, but leadership influence is generally limited to those directly under their charge. The SGM is a subject matter expert in his technical field, primary advisor on policy development, analytical reviewer of regulatory guidance and often fulfills the duties of the CSM in the incumbent's absence. SGMs also serve in non-staff and leadership positions such as Special Forces Team SGM, instructor at the Sergeants Major Academy or as the Senior Enlisted Advisor.

The Master/First Sergeant's Role

The 1SG is the senior NCO in companies, batteries and troops. The position of 1SG is similar to that of the CSM in importance, responsibility and prestige. As far back as the

Revolutionary War, 1SGs have enforced discipline, fostered loyalty and commitment in their Soldiers, maintained duty rosters, drilled, and made morning reports to their company commanders. Since today's 1SGs maintain daily contact with and are responsible for training and ensuring the health and welfare of the entire unit and Families, this position requires extraordinary leadership and professional competence. The MSG serves as the principal NCO in staff elements at battalion or higher levels. Although not charged with like leadership responsibilities of the 1SG, the MSG dispatches leadership and executes other duties with the same professionalism as the 1SG.

The Sergeant's First Class Role

a. While "Platoon Sergeant (PLT SGT)" is a duty position, not a rank, the PLT SGT is the primary assistant and advisor to the Platoon Leader, with the responsibility of training and caring for Soldiers. The PLT SGT helps the commander to train the Platoon Leader (PLT LDR) and in that regard has an enormous effect on how young officers perceive NCOs for the remainder of their career. The PLT SGT takes charge of the platoon in the absence of the PLT LDR. As the lowest level senior NCO involved in the company mission essential task list (METL), PLT SGT teach and train collective and individual tasks to Soldiers in squads, crews or small units.

b. SFC, may serve in a position subordinate to the PLT SGT or may serve as the NCOIC of the section with all the attendant responsibilities and duties of the PLT SGT. A PLT SGT or SFC generally has extensive military experience and can make informed decisions in the best interest of the mission and the Soldier.

The Staff Sergeant's Role

SSGs lead squads and sections and are a critical link in the NCO support channel. As a first line supervisor SSGs live and work with Soldiers every day and are responsible for their health, welfare, and safety. These leaders ensure that their Soldiers meet standards in personal appearance and teach them to maintain and account for their individual and unit equipment and property. The SSG enforces standards, develops, and trains Soldiers in MOS skills and unit missions. SSGs secondary role is to support the chain of command through the NCO Support Channel.

The Sergeant's Role

SGTs directly supervise Soldiers at the team level. The counseling, training, and care Sergeants provide will determine the outcome of battles won on the field and issues that develop during home station operations. SGTs are the first line leaders who have the most direct impact on Soldiers. SGTs also serve as the first line of the NCO Support Channel. NCO privileges are shown below. NCOs will—

(1) Function only in supervisory roles on work details and only as NCO in charge (NCOIC) of the guard on guard duty, except when temporary personnel shortages require the NCO to actively participate in the work detail.

(2) May be granted privileges by their respective commander in order to enhance the prestige of their enlisted troop leaders.

4.19. Precedence of relative grade, enlisted Soldiers. Precedence among enlisted Soldiers of the same grade in active military Service, to include retired enlisted Soldiers on AD, precedence or relative grade will be determined as follows:

a. According to Date of Rank. AR 600–20, 18 March 2008.

b. By length of active Federal Service in the Army when dates of rank are the same.

c. By length of total active Federal Service when a and b, above are the same.

d. By date of birth when a, b, and c, above are the same— older is more senior.

4.20. Duties, Responsibilities, and Authority

4.21. The three types of duties. As a NCO, you have duties and responsibilities that you must accomplish. These duties include:

a. **Specified Duties.** Specified duties are those related to jobs and positions, mainly military occupational specialty (MOS) related duties. Directives such as Army regulations (AR), Department of the Army (DA) general orders, the Uniform Code of Military Justice (UCMJ), Soldiers manuals, and MOS job descriptions specify the duties. For example, AR 600-20 says that NCOs must ensure that Soldiers are properly trained and maintain their personal appearance and cleanliness.

b. **Directed Duties.** Directed duties are those issued by superiors orally or in writing; these duties are not found in the unit's organizational charts. Directed duties include being in charge of quarters (CQ), serving as SGT of the guard, staff duty NCO, key control NCO, safety NCO, and crime prevention NCO.

c. **Implied Duties.** Implied duties often support specified duties, but in some cases they may not be related to the MOS job position. These duties may not be written, but implied in instructions. They are duties that improve the quality of the job and help keep the unit functioning at an optimum level. In most cases, these duties depend on individual initiative. They improve the work environment and motivate Soldiers to perform because they want to, not because they have to.

4.22. Types of Responsibility. Responsibility is being accountable for what you do or fail to do. Every Soldier has two responsibilities: Individual and Command responsibilities. The two most important responsibilities for a NCO are mission accomplishment and the welfare of their Soldiers. In addition to the two basic responsibilities, NCOs must:

- Maintain Discipline
- Maintain Property Accountability

- Train Soldiers

4.23. Types of Authority.

a. **Command Authority.** The key elements of command are authority and responsibility which are derived from policies, procedures, and precedents. Command is the authority a commander lawfully exercises over subordinates by virtue of rank or assignment. Command authority originates with the President and may be supplemented by law or regulation. Leading Soldiers includes the authority to organize, direct and, control assigned Soldiers so they accomplish assigned missions. It also includes authority to use assigned equipment and resources to accomplish your missions. Remember that this only applies to Soldiers and facilities in your unit.

b. **General Military Authority.** General military authority is the authority extended to all Soldiers to take action and act in the absence of a unit leader or other designated authority. It originates in oaths of office, laws, rank structure, traditions, and regulations. This broad-based authority also allows leaders to take appropriate corrective action whenever a member of any armed service, anywhere, commits an act involving a breach of good order and discipline. For example, if you see Soldiers in a brawl, you have the general military authority (and a duty) to stop the fight. This authority applies even if none of the Soldiers are in your unit.

Note: Only commanders are authorized to administer punishment as a result of due process. Leaders with general military authority are not authorized to punish, but can administer counseling or conduct corrective training. If a Soldier perceives that he/she is being punished, you may have overstepped your authority and may be subject to disciplinary action. An example of what is not corrective training is; "PVT Joe, you were late for formation, so you will do 500 push-ups starting at 1800."

(1) General military authority exists whether you are on or off duty, in uniform or in civilian attire and regardless of

location. For example, you are off duty, in civilian clothes and in the military clothing store and you see a Soldier in uniform with his headgear raised up and trousers un-bloused. You stop the Soldier immediately, identify yourself and ensure the Soldier makes the necessary corrections. If she/he refuses, stating you do not have the authority to tell her/him what to do because you are not her/his supervisor, you have the authority to take the Soldier into custody and turn her/him over to or report the incident to the Soldier's chain of command.

(2) You, as a NCO, have general military authority and the duty to enforce standards such as those outlined in AR 670-1, Wear and Appearance of Army Uniforms and Insignia. Your authority to enforce regulations is specified in AR 600-20 and if you neglect your duty, you subject yourself to disciplinary action.

4.24. Exercising Military Authority.

a. Military authority is exercised promptly, firmly, courteously and fairly. Commanders should consider administrative corrective measures before deciding to impose non-judicial punishment. Trial by court-martial is ordinarily inappropriate for minor offenses unless lesser forms of administering discipline would be ineffective.[55]

b. One of the most effective administrative corrective measures is extra training or instruction (including on-the-spot correction). For example, if Soldiers appear in an improper uniform, they are required to correct it immediately; if they do not maintain their housing area properly, they must correct the deficiency in a timely manner. If Soldiers have training deficiencies, they will be required to take extra training or instruction in subjects directly related to the shortcoming.

(1) The training, instruction, or correction given to a Soldier to correct deficiencies must be directly related to a deficiency. It must be oriented to improving the Soldier's performance in his or her problem area. Corrective measures

---

[55] U.S. Department of the Army, Military Justice, AR 27-10, 3 October 2011.

may be taken after normal duty hours. Such measures assume the nature of training or instruction, not punishment. Corrective training should continue only until the training deficiency is overcome. Authority to use it is part of the inherent powers of command.

(2) Care should be taken at all levels of command to ensure that training and instruction are not used in an oppressive manner to evade the procedural safeguards applied to imposing non-judicial punishment. Deficiencies satisfactorily corrected by means of training and instruction will not be noted in the official records of the Soldiers concerned.

(3) Military discipline is a state of order and obedience existing within a command. It involves the ready subordination of the will of the individual for the good of the group. Military discipline is an extension and application of habitual, but reasoned obedience that preserves initiative and functions even in the absence of the commander. Discipline is created within a command by instilling a sense of confidence and responsibility in each individual. When an individual knows he or she will be held accountable for their actions, then their individual discipline is heightened. Examples of discipline, manifested in individual behavior, are:

- Obedience to orders
- Military courtesy
- Maintenance of order
- Proper exercising military authority
- Observance of standards of conduct

4.25. Delegation of Authority. To meet the organization's goals, officers delegate authority to NCOs in the NCO Support Channel who, in turn, may further delegate that authority. Unless restricted by law, regulation, or a superior, leaders may delegate any or all of their authority to their subordinate leaders. However, such delegation must fall within the leader's scope of authority. Leaders cannot delegate authority they do not have and subordinate leaders may not assume authority that superiors do not have, cannot delegate, or have retained.

The task or duty to be performed limits the authority of the leader to whom it is assigned.

a. Consummate leaders obtain referent power by studying Army Regulations pertinent to their vocation. The foundation of understanding your authority starts with AR 600-20. It covers Army Command policy to include enlisted and NCO. authorities and responsibilities.

b. The Manual for Courts Martial (MCM, 2012) describes legal aspects of the authority of the NCO. It states in part that, "All commissioned officers, warrant officers and NCOs are authorized to stop quarrels, frays and disorders among persons subject to this chapter and to apprehend persons subject to this chapter who take part therein."[56] Severe penalties are imposed for violations such as disrespect, insubordination, or assault. No one expects you to be an expert on military law, but as a NCO you should know the definition of these words, apply the rules and procedures and be able to explain them to your Soldiers.

4.26. Soldier Conduct. As a leader you will be asked to evaluate an individual's character long before inappropriate behavior begins. In every case there are warning signs. NCOs are important to maintaining discipline in the Army. Army policies prescribed in this subparagraph should be considered together with the provisions of AR 27–10 and the MCM.

a. NCOs have the authority to apprehend any person subject to trial by court-martial under the MCM. NCOs not otherwise performing law enforcement duties should not apprehend a commissioned officer unless directed to do so by a commissioned officer.[57]

b. NCOs may be authorized by their commanders to order pretrial restraint of enlisted Soldiers of the commanding officer's command or enlisted Soldiers subject to the authority of that commanding officer per the MCM.

---

[56] U.S. Department of Defense, *Manual for Courts-Martial,* MCM 2012, 5 April 2012.
[57] Ibid.

c. NCOs do not have authority to impose non-judicial punishment on other enlisted Soldiers under the MCM (Part V, ART 15, UCMJ). However, the commander may authorize an NCO in the grade of SFC or above, provided such person is senior to the Soldier being notified, to deliver the DA Form 2627 (Record of Proceedings under UCMJ, ART. 15) and inform the Soldier of his or her rights. In cases of non-judicial punishment, the recommendation of a NCO should be sought and considered by unit commanders.

(1). As enlisted leaders of Soldiers, NCOs are essential to furthering the efficiency of the company, battery, or troop. This function includes preventing incidents that make it necessary to resort to trial by courts-martial or to impose non-judicial punishment. Thus, NCOs assist commanders in administering minor non-punitive corrective actions as found in AR 27–10 and Part V, paragraph 1g, of the MCM. "Non-punitive measures" are not "non-judicial punishment."[58]

(2). In taking corrective action with regard to subordinates, NCOs will be guided by and observe the principles listed in chapter 4.

d. Ensuring the proper conduct of Soldiers is a function of command. Commanders and leaders in the Army, whether on or off duty or in a leave status, will—

(1) Ensure all military personnel present a neat, Soldierly appearance.

(2) Take action consistent with Army regulation in any case where a Soldier's conduct violates good order and military discipline.

e. On public conveyances in the absence of military police, the person in charge of the conveyance will be asked to notify the nearest military police and arrange to have them, and if necessary, take custody of military personnel. In serious

---

[58] U.S. Department of the Army, Military Justice, AR 27-10, 3 October 2011.

situations, such as physical assault, the person in charge of the conveyance will be asked to stop at the first opportunity and request local police assistance. In all such cases, the local police will be advised to telephone (collect) the nearest Army post or Army headquarters.

f. When an offense endangering the reputation of the Army is committed elsewhere (not on a public conveyance) and military police are not available, civilian police will be requested to take appropriate action.

g. When military police are not present, the senior officer, warrant officer, or NCO present will obtain the Soldier's name, grade, social security number, organization, and station. The information and a statement of the circumstances will be sent to the Soldier's commanding officer immediately. If the Soldier is turned over to the civilian police, the above information will be sent to the civilian police for transmittal to the proper military authorities.

4.27. Authority to Apprehend. Rules for Courts- Martial Page II-17, Rule 302 (b) (1) and (2) (d)

- Military law Enforcement officials in the execution of law enforcement duties may apprehend.

- Commissioned, warrant, petty, and NCOs on active duty or inactive duty training may apprehend.

Note: Noncommissioned and petty officers, not otherwise performing law enforcement duties, should not apprehend a commissioned officer unless directed to do so by another commissioned officer, in order to prevent disgrace to the service, or to prevent the escape of one who has committed a serious offense.

4.28. How an apprehension may be made.

a. In general, an apprehension is made by clearly notifying the person to be apprehended that person is in custody. This notice should be given orally or in writing, but it may be implied by the circumstances.

b. Warrants. Neither warrants nor any other authorization shall be required for an apprehension under these rules except as required in subsection (e) (2) of this Rule 302.

c. Use of force. Any person authorized under these rules to make an apprehension may use such force and means as reasonably necessary under the circumstances to affect the apprehension.

d. Where an apprehension may be made. An apprehension may be made at any place, except in private dwellings, which does not include living areas in military barracks, vessels, aircraft, vehicles, tents, bunkers, field encampments, or similar places. Refer to Rule 302, (e) (A) (B) (C) and (D) for other exceptions when consent is required.

4.29. Maintenance of Order. Army and Marine Corps military police, Air Force security police, and members of the Navy and Coast Guard shore patrols are authorized and directed to apprehend Armed Forces members who commit offenses punishable under the UCMJ. Officers, warrant officers, NCOs, and petty officers of the Armed Forces are authorized and directed to quell all quarrels, frays, and disorders among persons subject to military law and to apprehend participants. Those exercising this authority should do so with judgment and tact. Personnel so apprehended will be returned to the jurisdiction of their respective Service as soon as practical. Confinement of females will be according to AR 190–47.

CHAPTER 5

Officer and NCO Relationships

Chapter 5.

Officer and NCO Relationships

5.0. Ref:

- ADP 6-22, Army Leadership
- ADRP 6-22, Army Leadership
- AR 600-20, Army Command Policy
- FM 6-22, Army Leadership: Competent, Confident, and Agile

5.1. Roles and Relationships. "When the Army speaks of Soldiers, it refers to officers, NCOs and enlisted men and women. The roles and responsibilities of Army leaders overlap and complement each other. Formal Army leaders come from three different categories: commissioned officers, Noncommissioned Officers, and Army Civilians. Collectively, these groups work toward a common goal and follow a shared value system."[59]

5.2. Army Officers and NCO relationship. Mutual trust and common goals are the two characteristics that enhance the relationship between Officers and NCOs. For instances, "NCOs have roles as trainers, mentors, communicators, and advisors. When junior officers first serve in the Army, their NCO helps to train and mold them. Doing so ensures Soldier safety while forming professional and personal bonds with the officers based on mutual trust and common goals"[60] NCOs are "the backbone of the Army"[61] and are the senior enlisted advisors who assist Commanders with knowledge and discipline for all enlisted matters.[62]

a. Every Soldier has a Sergeant. Officers are no exception. Platoon Sergeants, First Sergeants, Sergeants Major and

---

[59] U.S. Department of the Army, *Army Leadership*, ADRP 6-22, 1 August 2012.
[60] Ibid.
[61] Ibid.
[62] Ibid.

Command Sergeants Major at all levels serve as their respective officer's Sergeant.

b. An important part of your role as an NCO is how you relate to commissioned officers. To develop this working relationship, NCOs and officers must know the similarities of their respective duties and responsibilities.

c. Commissioned officers hold a commission from the President of the United States, which authorizes them to act as the President's representative in certain military matters. Laws, regulations, policies, and customs limit the duties and responsibilities of commissioned officers, NCOs, and government officials. As the President's representatives, commissioned officers carry out the orders of the Commander-in-Chief as handed down through the chain of command. In carrying out orders, commissioned officers get considerable help, assistance and advice from NCOs.

d. Noncommissioned Officers obtain their authorities as agents of the Secretary of the Army outlined in Army regulations. They support the command authority of commissioned officers. As the Secretary of the Army's representatives, Noncommissioned Officers carry out the orders of Commander-in-Chief through the chain of command. In carrying out orders, Noncommissioned Officers provide support, assistance and advice to officers.

e. A commissioned officer:

- Commands and establishes policy, plans, and programs the work of the Army.
- Must be technically and tactically proficient in his/her MOS and that of the organization.
- Concentrates on collective training to enable the unit to accomplish the mission
- Is primarily involved with unit operations, training and related activities
- Pays particular attention to the standards of performance, training and professional development of officers as well as NCOs

- Creates conditions – make time and other resources available – so the NCO can do the job.
- Supports the NCO.

f. A Noncommissioned Officer:

- Conducts the daily business of the Army within established orders, directives and policies.
- Focuses on individual training, which develops the capability to accomplish the mission
- Is primarily involved with training and leading Soldiers and teams
- Ensures each subordinate team, NCO and Soldier is prepared to function as an effective unit and each member is well trained, highly motivated, ready and functioning.
- Concentrates on standards of performance, training, and professional development of NCOs and enlisted Soldiers.
- Follows orders of officers and NCOs in the support channel.
- Gets the job done.
- Must be an effective trainer.
- Must develop foresight and keep standards high.
- Must possess the courage to act.
- Must be a subject matter expert in their MOS, able to provide clear and concise input relevant to MDMP.

g. The success of the U.S. Army is directly related to the quality of the professional relationships between its officers and Noncommissioned Officers. The officer/ NCO team forms the cornerstone of our Army and when the bond is formed, it can have the single most important impact on unit effectiveness and efficiency. Conversely, if the bond is broken, it can have a devastating impact on morale, esprit de corps, readiness and mission accomplishment.

h. AR 600-20 defines command authority and what is NCO support, but the officer/NCO relationship is not created by a strict set of rules, policy or procedures, it is based on mutual respect, communication, trust, commitment and devotion. The officer/NCO relationship benefits officers at all levels, but especially junior officers. The relationship established during junior officer development will have a lasting impact on those officer's opinions, respect for, good will, and confidence in Noncommissioned Officers for the rest of their careers.

i. Noncommissioned Officers accept as an unwritten duty, the responsibility to instruct and develop second lieutenants, but it is the company commander's responsibility to train lieutenants and the battalion commander who is the driving force behind the training of lieutenants.

j. There are several critical officer/NCO relationships that form the bond as a team: Platoon Sergeant/Platoon Leader, First Sergeant/Company Commander, Battalion CSM/Battalion Commander, and CSM/Brigade Commander. Both NCOs and officers have expectations of each other that form the foundation of a strong relationship.

k. What should the NCO expect of an officer? The NCO can expect the officer to:

- Have personal integrity and high morals.
- Maintain a high state of appearance- be a standard-bearer.
- Be fair, be consistent, and have dignity.
- Be compassionate and understanding- do not be aloof to the issues and problems of Soldiers.
- Have courage in the face of danger.
- Have courage of convictions and stand up for what is right, even though it might be hard.
- Not expose themselves or Soldiers to unnecessary risk.
- Be accountable for their own actions and the actions of their Soldiers.
- Endure hardships equal to the hardships experienced by Soldiers.

l. What should the officer expect of an NCO? The officer can expect the NCO to:

- Be loyal to the officer's position.
- Be devoted to the cause of national defense.
- Have admiration for the officer's honest effort.
- Have endurance that matches officers.
- Have motivation that matches officers.
- Have intestinal fortitude and courage.
- Have a strong desire to achieve goals that matches officers.
- Have a strong spirituality, love of country, and a love of duty that matches officers.
- Endure hardships equal to the hardships experienced by Soldiers.
- Master expertise in Army programs that supports the needs of Soldiers and their Families.

m. What should a Platoon Sergeant expect of their Platoon Leader? The PSG can expect the PL to:

- Let the NCOs handle the problems of the platoon.
- Be pleasant and approachable.
- Look every inch an officer in act, word and deed.
- Show maturity.
- Recognize the imbalance of experience.
- Ensure the PL and PSG begin with common goals.
- Communicate. Good communication does not happen all by itself. Talk, talk, talk and listen, listen, listen.
- Counsel. The PL and PSG must work together to establish realistic, recognizable standards and after counseling a Soldier, communicate the results to each other.
- Give the company commander and the First Sergeant a perspective of how he/she is getting along with the PSG.
- Lead squad leaders and depend on NCOs to directly lead individual Soldiers.

n. What should Platoon Leader expect of their Platoon Sergeant? The PL can expect the PSG to:

- Understand the inherent responsibility to coach and counsel PL's to develop their competence, character, and commitment in the performance of their duty. Developing junior officers is a PSG's responsibility.

- Recognize the imbalance of experience.
- Demonstrate tact and diplomacy with the PL.
- Offer advice, but execute orders.
- Incorporate the PL into the team he/she has to lead.
- Mold, guide, and educate the PL to the subtleties of Army life.
- Share knowledge and experience.
- Train and correct the PL when needed.
- Show a genuine concern that the PL is learning the right way instead of the easy way.
- Do not undermine or destroy the PLs credibility (Remember that order/counter-order creates disorder).
- Set the example for the PL through military bearing and consistent demonstration of character, competence, and commitment to the mission, Soldiers, and their families.
- When the PL makes a mistake, make sure they learn from those mistakes, if repeated; provide firm, pointed instruction to keep it from being habitual.
- Give the company commander and the First Sergeant a perspective on how the PL is doing.
- Give the PL the PSG's view on particular matters before the PL discusses with the company commander.

o. What should the First Sergeant expect of the company commander? The First Sergeant can expect the company commander to:

- Possess the same qualities expected of all officers.
- Have a rapport with the battalion CSM.
- Administer fair and impartial justice.
- Take responsibility for their actions and those of the unit and Soldiers.
- Seek the First Sergeant's advice.
- Never belittle or undermine the First Sergeant and respect the position or the First Sergeant's authority.

p. What should the company commander expect of the First Sergeant? The company commander can expect the 1SG to:

- Possess the same qualities expected of all NCOs.
- Maintain discipline.
- Train, educate, and share experiences with both the commander and Soldiers.
- Be loyal to the commander's position.
- Develop and agree on the goals, standards and objectives of the company.
- Have mutual trust and respect for each other.
- Know the commander's strengths and weaknesses.
- Know their responsibilities as defined in AR 600-20.
- Possess a strong sense of duty.
- Ready the company for any mission.
- Be the standard-bearer in appearance, morals, ethics, values, competences, and commitment.
- Be an advisor, but execute orders.
- Be the subject matter expert in Army programs that best supports the needs of Soldiers and their Families.

q. What should the Command Sergeant Major expect of the battalion/brigade commander? The CSM can expect the BC to:

- Possess the same qualities expected of all officers.
- Seek advice and share views.
- Have open communications with the CSM.
- Be fair and impartial.
- Inspire leaders and Soldiers.
- Understand each other and how they will function together as a team.
- Harness the CSM's talents.
- Do not limit the CSM's duties or responsibilities.
- Resource the CSM as an enlisted extension of the BC.
- Learn to know the CSM's feelings about any given subject.
- Give the position of CSM the respect it is due.

- Empower the NCO Support Channel to solve problems at the lowest level.
- Ensure there is no one in the chain of command that comes between the BC and the CSM.

r. What should the battalion/brigade commander expect of the CSM? Commanders can expect the CSM to:

- Possess the same qualities expected of the BC.
- Share views.
- Visit Soldiers on the ground and get their perspective. Inspect and check where Soldiers are.
- Inspire leaders and Soldiers.
- Be a leader of presence and character.
- Manage the organization's sponsorship program.
- Manage processes and procedures.
- Be an expert in customs, courtesies, traditions and ceremonies.
- Understand each other and how they will function together as a team.
- Be a reliable, trusted confidant.
- Have honest and candid communications and be able to disagree without being disrespectful.
- Ensure there is no one in the chain of command that comes between the team.
- Have direct access and be accountable to the BC.
- After obtaining advice and making a decision, the CSM supports those decisions.
- Ensure the CSM provides advice to company commander/First Sergeant relationships, CSM/1SG relationships, CSM/company commander relationships vs. the company commander/battalion staff relationship.
- Ensure the CSM is responsible for assigning incoming Noncommissioned Officers.
- Be impartial and be objective.
- Conduct inspections, check training, sit as president of promotion boards, and be a part of the reenlistment program.
- Work closely with the EO/EEO/Chaplain and SHARP Advocate.

- Be the subject matter expert on all Army programs to support Soldiers and their Families.
- Be the most experienced trainer in the organization.
- Learn to know the commander's feelings about any given subject.

5.3. Army Civilians and NCO relationship. Army Civilians are skilled personnel dedicated to serving the nation as an integral part of the Army team. They provide mission-essential capability, stability, and continuity during war and peace to support Soldiers. The Army civilian and NCO relationship enables the Army to continue to accomplished its mission through performance, while contributing to the overall organizational goals.[63]

---

[63] U.S. Department of the Army, *Army Leadership*, ADRP 6-22, 10 September 2012.

CHAPTER 6

TRAINING

Chapter 6

Training.

6.0. Ref.

- ADP 7-0, Training Units and Developing Leaders
- ADRP 7-0, Training Units and Developing Leaders
- AR 350-1, Army Training and Leader Development
- AR 623-3, Evaluation Reporting System
- AR 735-5, Property Accountability Policies
- FM 7-15, Army Universal Task List

6.1. Unit Training Management. "Training is the primary focus of a unit when not deployed. It requires the same level of detail, intensity, and focus that a unit applies to deployed operations. The operations process provides a common framework for units to plan, prepare, execute, and assess training and to integrate leader development into training plans. Battalions and higher units use the military decision-making process to develop unit training plans; companies use troop leading procedures to develop unit training plans."[64]

---

[64] U.S. Department of the Army, *Training Units and Developing Leaders*, ADP 7-0, 23 August 2012.

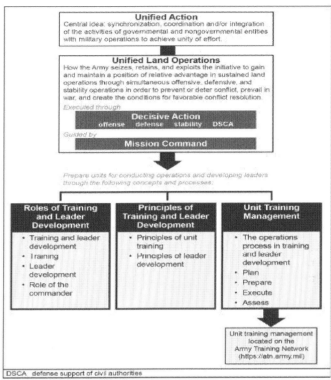

Figure 25. Unit Training and Leader Development Underlying Logic

6.2. The Training and Leader Development.

a. The Army prepares to conduct unified land operations in support of unified action by conducting tough, realistic, standards-based, performance-oriented training.

(1) NCOs are the primary trainers of enlisted Soldiers, crews, and small teams. NCOs take broad guidance from their leaders: indentify the necessary tasks, standards, and resources; and then plan, prepare, execute, and assess training.

(2) NCOs ensure Soldiers demonstrate proficiency in their individual military occupational specialty skills, warrior tasks, and battle drills.

(3) NCOs instill in Soldiers discipline, resiliency, the Warrior Ethos, and Army Values.

(4) NCOs provide feedback on task proficiency and the quality of training.

(5) Leaders allot sufficient time and resources, and empower NCOs to plan, prepare, execute, and assess training.

b. Unit Training Management is an essential part of a unit's leader development program. Sergeant's Time Training (STT), the process used by Army leaders to identify training requirements, is a common approach to NCO –led training events. NCOs conduct STT to standard not to time.

6.3. A NCO's Principal Duty is to Train.

a. Understanding the importance of training management will assist the NCO in reducing training time and training effort while increasing training effectiveness and training sustainment. By understanding the linkage between institutional, organizational, and self-development, the trainer will be better prepared in developing Soldiers for combat.

b. Training is managed at units through the Digital Training Management System (DTMS). You must be registered as a user in your unit to access the system. The following link provides more detailed information on its use and importance.
http://www.army.mil/aps/08/information_papers/transfor m/Digital_Training_Management_System.html

c. Resident training base and distributed learning is managed at units and institutions through the Army Training Requirements and Resources System (ATRRS). The following link provides more information on its use.
https://www.atrrs.army.mil/

d. Training plans and management of life-long learning is managed by the Army through the Army Career Tracker. The following link provides more details and information on the ACT and its uses.

http://www.civiliantraining.army.mil/Pages/Army-Career-Tracker.aspx

e. The Army Training Network (ATN) provides digital training resources in support of training management. ATN provides direct enablers to enhance unit training management (UTM) such as Army Universe Task List (AUTL), Combined Arms Training Strategies (CATS), Digital Training Management System (DTMS), Mission Essential Task Lists (METL), Training Aids Devices, Simulators, and Simulations (TADSS), training videos, and other training related products. The website to access the ATN is:
https://atn.army.mil/cac2/atn/

f. The NCO Corner, located in the ATN Website provides immediate access to other support capabilities such as the Army Learning Management System (ALMS), the Army Career Tracker, My Army benefits and other pertinent issues related to NCO and Soldier development and care.

6.4. NCO Development. The three domains found in the Army's leadership development model (Figure 26) are essential to the NCOs development through training, experience and education.

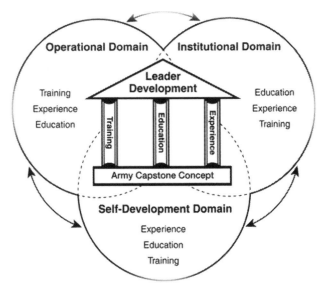

Figure 26. The Army's Leader Development Model

a. **The institutional domain.** The institutional training domain is a composite of the Army School system that includes Army centers/schools that provide initial military training, reclassification training, functional training, and Army professional development training and education for Soldiers, military leaders, and Army civilians. Army schools ensure Soldiers, leaders, and Army civilians can perform critical tasks to prescribed standard. The institutional training domain also provides training support products, information, and materials needed by individuals for self-development and by unit leaders in the operational domain to accomplish training and mission rehearsal/assessment. Institutional training supports every Soldier and Army Civilian in the force throughout their Army career.[65]

(1) MOS specific education. MOS training is the Army's professional development Training and Education Program. Selection of Soldiers for leader training and

---

[65] U.S. Department of the Army, *Army Training and Leader Development*, AR 350-1, 19 August 2014.

education is linked to promotion, future assignments, and career management conducted in specific MOS education in the NCOES.

(2) Mandatory/Directed Training. Consist of HQ DA-selected general subject areas in which individual Soldiers and Army Civilians must be proficient to perform satisfactorily in a military organization.[66]

(3) Functional Training. Selection for functional and specialty courses is linked to operational assignments that require skills and knowledge not trained in NCOES.

b. **The operational domain.** The operational domain encompasses training activities that individuals, units and organizations undertake. These activities include training conducted at home station, during major training events (to include Joint exercises) at combat training centers and other locations (to include mobilization centers), and while operationally deployed. Unit leaders are responsible for the proficiency of their subordinates (Soldiers and Army civilians), subordinate leaders, teams/crews, and the unit as a whole in relation to unit readiness to prepare to conduct its wartime mission.

(1) Mandatory/Directed Training. As with institution training requirements, mandatory subjects are required to be taught at organizations and consist of HQDA-selected general subject areas in which individual Soldiers and Army Civilians must be proficient to perform satisfactorily in a military organization. The Army e-learning program provides an efficient and effective program to accomplish workforce training in information technology, information assurance, foreign languages, and selected mandatory training requirements.[67]

(2) Individual Tasks. Focuses training on supporting tasks necessary for collective training. The Army's Universal

---

[66] U.S. Department of the Army, *Army Training and Leader Development*, AR 350-1, 19 August 2014.
[67] Ibid.

Task List provides the composition of tasks essential to individual development outside MOS professional development training and education, but supports collective and mission essential critical tasks.[68]

(3) Collective Training. Focuses training on mission essential tasks and supporting critical tasks.

c. **Self-development domain.**

(1) The Army defines self-development as planned, goal-oriented learning that reinforces and expands the depth and breadth of an individual's knowledge base, self-awareness, and situational awareness. Self-development will complement what has been learned in the classroom and on the job, enhance professional competence, and help meet personal objectives. There are three types of self-development:

(a) Structured Self-Development. Required learning that continues throughout a career and that is closely linked to and synchronized with classroom and on-the-job learning. SSD is to bridge the operational and institutional domains and set the conditions for continuous growth. SSD builds knowledge and skills through a defined sequence of learning approaches involving formal education and experiential learning. The implementation of SSD will improve Army readiness by integrating self-development into a life-long learning approach that fosters continuous learning. SSD is a prerequisite for attendance to NCOES courses.[69]

- SSD I – Required for Warrior Leader Course.
- SSD II – Required for Advanced Leader Course.
- SSD III – Required for Senior Leader Course.
- SSD IV – Required for Sergeants Major Course.
- SSD V – Required for Nominative and Joint Assignments.

---

[68] U.S. Department of the Army, *The Army Universal Task List*, FM 7-15, 27 February 2009.

[69] U.S. Department of the Army, *Army Training and Leader Development*, AR 350-1, 19 August 2014.

NOTE: Consistent with recent guidance; successful completion of a prerequisite level of SSD is required prior to making a reservation in the Army Training Requirements And Resources Systems (ATRRS) for the corresponding level of NCOES.[70]

(b) Guided self-development. Recommended but optional learning that will help keep personnel prepared for changing technical, functional, and leadership responsibilities throughout their career.

(c) Personal self-development. Self-initiated learning where the individual defines the objective, pace, and process.

(2) The self-development training domain recognizes that Army service requires continuous, life-long learning and that structured training activities in Army schools and in operational units often will not meet every individual's need for content or time.

(a) Self-development enables individuals to pursue personal and professional development goals. Leaders help subordinates identify areas where self-development will improve performance of current assignment and areas that will prepare them for future career assignments.

(b) Army schools provide training and education products that can be used for self-development. Self-development focuses on maximizing strengths, overcoming weaknesses, and achieving individual development goals.

(c) All Soldiers and Army civilians must accept their personal responsibility to develop, grow, and commit to professional excellence. The Soldier or Army civilian must commit to a lifetime of professional and personal growth focused on staying at the cutting edge of their profession.

---

[70] U.S. Department of the Army, *Transition Of Advanced Leader Course Common Core To Structured Self Development Level 2*, ALARACT 126/2014, 15 May 2014.

(d) The success of self-development is tied to regular self-assessment and performance feedback. Individuals must regularly conduct honest assessments of their strengths and weaknesses. First line leaders must regularly provide feedback on performance and assist individuals in establishing/ refining an individual development action plan to guide performance improvement.

(e) A second critical component of success is ensuring uniform access to current learning materials and programs. The self-development domain must be a robust component of the professional development model (PDM) providing every Soldier and Army civilian clear understanding of what success looks like. The PDM can be found at **https://atiam.train.army.mil/Soldier Portal/**.

6.5. NCO Education System (NCOES). The goal of NCO training and education is to prepare NCOs to lead and train Soldiers who work and fight under their supervision and to assist their leaders to execute unit missions.

a. NCOES is linked to promotion to SGT, SSG, SFC, and SGM.

b. NCOES provides the NCO with progressive and sequential leader, technical, and tactical training relevant to the duties, responsibilities and missions they will perform in operational units.

c. Life Long Learning. The NCOES is geared towards more training in the early stages of Soldier and leader development. As leaders progress, less emphasis is placed on training and more emphasis is placed on education.

d. Training is focused more on a "step-by-step" list of what needs to be done to accomplish the skill being learned and accomplished when the trainee can reiterate the right answers and/or demonstrate the "approved way" of doing something. Training is specific, has a definite goal and a time, and requires a show of proficiency.

e. Education is broader than training. It prepares learners to be critical and analytical thinkers for effective problem solving by facilitating the learning of principles, concepts, rules, facts, and associated skills and values/attitudes. Education aims to develop a NCO's understanding, abilities to synthesize information, and enhance skills within and beyond their role.

6.6. NCO Development Program (NCODP). Essential to leader development is reinforcement through unit professional development programs. NCODP sessions are tailored to unique unit requirements and support the commander's leader training and leader development program. NCO development is achieved through a progressive sequence of local and Army-level education, unit and individual training, and assignments of increasing scope and responsibility. The NCO Corps has a small population of senior NCO who serve in positions at strategic assignments. While small in number, applicable regulations address preparing and identifying the right talent to fill these high-visibility positions. These include SGM of the Army, senior NCO in Army commands and combatant commands, Army Staff, and congressional liaison positions. These are all referred to as nominative SGM positions.

a. When considering leader development in units, assignments of increasing scope and responsibility linked to broadening assignments is key to career management and development, such as progressing from Squad Leader to Platoon Sergeant and taking assignment and serving as an instructor or recruiter.

b. ADRP 7-0 describes unit leader training and leader development programs. DA Pam 350-58 describes the Army's approach for development and implementation of leader development programs. The Army's Leader Development Strategy (ALDS 2013) provides the direction to help the Army re-balance three crucial leader development components; training, education and experience.

c. Leader training ensures leaders can perform current assigned responsibilities.

d. Leader development prepares them for increased levels of responsibilities for the future.

6.7. Command Supply Discipline Program (CSDP). CSDP is a commander's program. However, leaders must enforce and implement effective programs to ensure resources are not subject to fraud, waste and abuse. CSDP is an individual, supervisory and managerial responsibility that standardizes supply discipline throughout the Army. An essential characteristic of our responsibility is stewardship of our profession. A key role and responsibility of a NCO is accountability and readiness of property and stewardship of Army resources. Besides the readiness of Soldiers, the single most important combat multiplier is equipment readiness and serviceability. The purpose of CSDP is:

a. Establish supply discipline as regulatory guidance.

b. Standardize supply discipline requirements.

c. Provide responsible personnel with a single listing of all existing supply discipline requirements.

d. Make the Army more efficient regarding time spent monitoring subordinates' actions.

6.8. Property Accountability.

a. All persons entrusted with government property are responsible for its proper care, custody, safekeeping and disposition.

b. Army Property will not be used for any private purpose except as authorized by HQDA.

c. No government property will be sold, given as a gift, loaned, exchanged or otherwise disposed of.

d. Giving or accepting an issue document, hand receipt, or other forms of receipt to cover articles that are missing or appear to be missing is prohibited.

e. Military members or Civilian employees of the Army who occupy government quarters, or issued furnishings for use in government quarters must properly care for such property.

6.9. Accounting for Army Property.

a. All property (including historical artifacts, art, flags, organizational property, and associated items) acquired by the Army from any source, whether bought, acquired, or donated must be accounted for.

b. Accounting will be continuous from the time of acquisition, until the ultimate consumption or disposal of the property occurs.

6.10. Item Classification.

a. **Nonexpendable Property.** Nonexpendable Property is personal property that is not consumed in use and that retains its original identity during the period of use. Nonexpendable property requires formal property book accountability at the user level. Examples are major end items such as HMMWV, weapons, tents, tarpaulins, flags, pennants, national flags.

b. **Expendable Property.** Expendable property is property consumed in use, or loses its identity in use. It includes items not consumed in use, with a unit cost of less than $500. Expendable items are items that require no formal accounting after issue to the user level. Example: oil, paint, fuel, or cleaning material.

c. **Durable Property.** Durable property is personal property that is not consumed in use and that does not require property book accountability but, because of its unique characteristics, requires control when issued to the user. It is property that requires control at the user level using hand-receipt procedures or managed using inventory lists. Example: computers, laptops, and software.

6.11. Inventories of personal property. All on-hand property carried on property book records and/or hand receipt records at the user's level will have a complete physical

inventory upon change of the primary hand receipt holder or accomplished annually, whichever occurs first.

6.12. CSDP Responsibility. Responsibility is the obligation of an individual to ensure government property and funds entrusted to their possession are properly used and cared for, and that proper custody, safekeeping and disposition are provided. Types of responsibility are:

a. Command responsibility. Commanders are obligated to ensure all government property is properly used, care for, safeguarded, in proper custody, and disposed of. Command responsibility is inherent in command and cannot be delegated.

b. Supervisory responsibility. Obligates supervisors to ensure all government property is properly used, cared for, safeguarded, in proper custody, and disposed of. It is inherent in all supervisory positions and is not contingent upon signed receipts or responsibility statements and cannot be delegated.

c. Direct responsibility. Obligates a person to ensure all government property is properly used, cared for, safeguarded, in proper custody, and disposed of. Direct responsibility results from assignment as an accountable officer or acceptance of the property on hand receipt from an accountable officer.

d. Custodial responsibility. Obligates an individual for property in storage, awaiting issue, or turn-in to exercise reasonable and prudent actions to property care for and ensure proper custody, safeguarding, and disposition of the property is provided.

e. Personal responsibility. Obligates a person to exercise reasonable and prudent actions to properly use, care for, safeguard, and dispose of all government property issued for, acquired for, or converted to a person's exclusive use, with or without a receipt.

CHAPTER 7

ARMY PROGRAMS

Chapter 7.

Army Programs

7.0. Ref.

- AR 215-1, Military Morale, Welfare, and Recreation Programs and Non-appropriated Fund Instrumentalities
- AR 350-1, Army Training and Leader Development
- AR 385-10, The Army Safety Program
- AR 600-8-8 – The Total Army Sponsorship Program (TASP)
- AR 600-20, Army Command Policy, Chap 6
- AR 600-85, The Army Substance Abuse Program
- AR 601-280, Army Retention Program
- AR 608-1, Army Community Service Center
- AR 608-18, The Army Family Advocacy Program
- AR 608-75, Exceptional Family Member Program
- AR 621-5, Army Continuing Education System
- AR 930-4, Army Emergency Relief
- AR 930-5, American National Red Cross Service Program and Army Utilization
- Army Directive 2011-19, Expedited Transfer or Reassignment Procedures for Victims of Sexual Assault
- Army Reserve Family Programs, **http://www.arfp.org/programs.html**
- DA Pam 385-10, The Army Safety Program
- Public Law
- Total Army Strong Program
- United States Army Combined Arms Center, **http://usacac.army.mil/**
- USAR 608-1, Army Reserve Family Programs

7.1. NCOs are charged with taking care of their Soldiers. The scope of this charge includes personal and professional issues. A primary way to do so is to have an in-depth knowledge, and use of the programs the Army has available to assist with the process. These programs were developed with the Soldiers and their Families' needs in mind. The list below provides available programs and applicable references to assist leaders in advising and counseling Soldiers. You may also follow the link below to the Army One Source website for additional information. https://www.myarmyonesource.com. Publications listed below can be found at http://www.apd.army.mil/ProductMap.asp.

7.2. American Red Cross. The American Red Cross (ARC) exists to provide compassionate care to those in need. Our network of generous donors, volunteers and employees share a mission of preventing and relieving suffering, at home and around the world through five key service areas: disaster relief, supporting America's military Families, lifesaving blood, health and safety services, and international services. The American Red Cross directly supports Soldiers' needs such as emergency notifications through Red Cross Messages. More information can be found at http://www.redcross.org/ and http://www.apd.army.mil/pdffiles/r930_5.pdf

7.3. Army Community Service. The purpose of Army Community Service (ACS) is to facilitate a commander's ability to provide comprehensive, coordinated, and responsive services that support readiness of Soldiers, civilian employees and their Families. Families can seek ACS support through family Assistance Centers, Family Readiness Groups, Rear Detachment, and Family Readiness Support Assistants. Soldier and Family readiness is supported through the Family Advocacy Program, Victim Advocacy Program, SHARP, Sexual Assault Prevention and Response Program (SAPR), Exceptional Family Member Program (EFMP), Transitional Compensation assistance, and New Parent Support Program. In addition, ACS provides support for relocation readiness, employment readiness, financial readiness, volunteer programs, Survivor outreach Services, and the Army OneSource initiative. More information on ACS can be found at http://www.apd.army.mil/pdffiles/r608_1.pdf or http://myarmybenefits.us.army.mil/Home/Benefit_Library /Federal_Benefits_Page/Army_Community_Service_(ACS).html?serv=243

7.4. Army Continuing Education System. The Army Continuing Education System (ACES) provides programs and services to promote lifelong learning opportunities and to sharpen the competitive edge of the Army. ACES improve combat readiness and resilience through flexible and relevant education programs, services and systems in support of the Total Army Family. ACES provides Army credentialing opportunities, tests of general education development (GED), high school completion program (HSCP), and FAST program that includes Basic Skills Education Program (BSEP), GT improvement, GED test preparation, reading skill development, testing services, transcript service, education counseling, and preparation for college. More information on ACES can be found at **http://www.apd.army.mil/pdffiles/r621_5.pdf** or **http://myarmybenefits.us.army.mil/Home/Benefit_Library /Federal_Bene-fits_Page/Army_Continuing_Education_System.html?serv =228**

7.5. Army Emergency Relief. Army Emergency Relief (AER) is the Army's own emergency financial assistance organization and is dedicated to "Helping the Army Take Care of Its Own."[71] AER provides commanders a valuable asset in accomplishing their basic command responsibility for the morale and welfare of Soldiers. As a leader, it is important to understand the rules and resources available to you and Soldiers. Personnel eligible for AER are;

a. Soldiers on extended active duty and their dependents.

b. Reserve Component Soldiers (ARNG and USAR) on continuous AD orders for more than 30 consecutive days and their dependents. (This applies to Soldiers on AD for training (ADT) and serving under various sections of title 10, United States Code.

c. Soldiers retired from AD for longevity, retired by reason of physical disability, or retired at age 60 under Section 1331, Title 10, United States Code (10 USC 1331) and their dependents.

d. Surviving spouses and orphans of eligible Soldiers who died while on AD or after they were retired as identified in *c*

---

[71] Army Emergency Relief Home Page, http://www.aerhq.org.

above. More information on AER can be found at
http://www.aerhq.org/dnn563/ or
http://www.apd.army.mil/jw2/xmldemo/r930_4/main.asp

7.6. Army Family Action Plan (AFAP). AFAP creates an
information loop between the global Army family and
leadership. AFAP is the Army's grassroots process to identify
and elevate the most significant quality of life issues affecting
Soldiers of all components, retirees, Department of Army
(DA) civilians, and Families to senior leaders for action.
Leaders, Soldiers and family members are integral to
providing information to improve standards of living and
institute information and support program. More information
on AFAP can be found at
https://www.myarmyonesource.com/familyprogramsandse
rvices/familyprograms/armyfamilyactionplan/default.aspx
or
http://www.armymwr.com/UserFiles/file/BOSS/AFAP.pdf

7.7. Army Family Advocacy Program. The US Army
Family Advocacy Program is dedicated to the prevention,
education, prompt reporting, investigation, intervention and
treatment of spouse and child abuse. The program provides a
variety of services to Soldiers and Families to enhance their
relationship skills and improve their quality of life. More
information on the Family Advocacy Program can be found at
https://www.myarmyonesource.com/FamilyProgramsandS
ervices/FamilyPrograms/FamilyAdvocacyProgram/default
.aspx or
http://www.apd.army.mil/pdffiles/r608_18.pdf

7.8. Army Family Readiness Group. Army Family
Readiness Group (AFRG) provides all the functionality of a
traditional FRG in an ad-hoc and on-line setting to meet the
needs of geographically dispersed Units and Families across
all components of the Army. More information on AFRG can
be found at
https://www.armyfrg.org/skins/frg/display.aspx?moduleid
=8cde2e88-3052-448c-893d-
d0b4b14b31c4&CategoryID=1e9b3feb-985e-4dd2-b5b2-
db37a8aa0d63&ObjectID=bf2bcd52-fa60-4868-aa94-
f9f4a70a4e98

7.9. Army Family Team Building. Army Family Team Building (AFTB) empowers individuals, maximizing their personal growth and professional development through specialized training, transforming our community into a resilient and strong foundation meeting today's military mission. More information on AFTB can be found at **https://www.myarmyonesource.com/FamilyProgramsandS ervices/FamilyPrograms/ArmyFamilyTeamBuilding/defau lt.aspx**

7.10. Army Retention Program. Personnel readiness is a responsibility of command. DA Policy is that only those Soldiers who have maintained a record of acceptable performance will be offered the privilege of reenlisting within the Active Army. Reenlistment is the Army's equivalent of the quality management program, but at the organizational level. It is a leader responsibility to ensure only the best-qualified Soldiers are reenlisted. The Army's ranks and formations are strengthened by evaluating the whole Soldier when determining their future service. More information on the Army retention program can be found at **http://www.apd.army.mil/pdffiles/r601_280.pdf**

7.11. Army Safety Program. No other program has more impact on Soldier Readiness. It is every Soldier and Army Civilian's responsibility to stop unsafe acts by being responsible for accident prevention and applying risk management. This is also accomplished by compliance with the Army Safety Regulation, safety regulations work practices, standing operating procedures and by using all necessary personal protective equipment (PPE). It is also required to report Army accidents and hazards in the workplace and to employ risk management to manage risk. Safety goals will support overall command objectives by helping keep personnel safe and ready for duty. Leaders are safety officers who mitigate risk. More information on the Army Safety Program can be found at **https://safety.army.mil/**

7.12. Army Substance Abuse Program. The Army Substance Abuse Program (ASAP) mission is to strengthen the overall fitness and effectiveness of the Army's workforce, to conserve manpower, and to enhance the combat readiness of Soldiers. More information on ACSAP can be found at http://acsap.army.mil/sso/pages/index.jsp or http://armypubs.army.mil/epubs/pdf/r385_10.pdf or http://armypubs.army.mil/epubs/pdf/r600_85.pdf

7.13. Army World Class Athlete Program. The Army World Class Athlete Program (WCAP) provides support and training for outstanding Soldier-athletes to help them compete and succeed in national and international competitions leading to Olympic and Paralympic Games, while maintaining a professional military career. More information on WCAP can be found at http://www.thearmywcap.com/

7.14. Better Opportunities for Single Soldiers (BOSS). The BOSS is a quality of life program that addresses single Soldier issues and initiatives. The Better Opportunities for Single Soldiers program enhances the morale and welfare of single Soldiers, increase retention and sustain combat readiness. BOSS is the collective voice of single Soldiers through the chain of command, which serves as a tool for commanders to gauge the morale of single Soldiers regarding quality of life issues. More information on the BOSS program can be found at http://www.armymwr.com/recleisure/single/boss.aspx

7.15. Center for the Army Profession and Ethic. The Center for the Army Profession and Ethic (CAPE) serves as the proponent for the Army Profession, the Army Ethic and Character Development of Army Professionals to reinforce Trust within the profession and with the American people, which is the foundation for successful accomplishment of the Army Mission, consistent with the Army Ethic. More information on CAPE can be found at http://cape.army.mil/mission.php

7.16. Child, Youth, and School Services. Child, Youth and School (CYS) Services consists of four services; Child Development Services (CDS); School Age Services (SAS), Youth Services (YS) and CYSS Liaison, Education, and Outreach Services (CLEOS). CYS recognizes the challenges of our Soldiers and their Families. By offering quality programs for children, youth and students, CYS supports the Army Family Covenant by reducing the conflict between mission readiness and parental responsibility. More information on the CYS program can be found at http://www.armymwr.com/family/childandyouth/

7.17. Comprehensive Soldier and Family Fitness Program/MRT–
a. The Comprehensive Soldier and Family Fitness (CSF2) is designed to build resilience and enhance performance of the Army Family- Soldiers, Families and Army Civilians. CSF2 provides hands-on training and self-development tools so that members of the Army Family are better able to cope with adversity, perform better in stressful situations and thrive in the military and civilian sector and to meet a wide range of operational demands. The program emphasizes social, physical, family, spiritual, and emotional fitness.
b. Master Resiliency Training, part of the Comprehensive Soldier and Family Fitness program, offers strength-based, positive psychology tools to aid Soldiers, leaders, and Families in their ability to grow and strive in the face of challenges and bounce back from adversity. Training and information is targeted to all phases of the Soldier deployment cycle, Soldier life cycle, and Soldier support system. More information on the CSF2 program can be found at http://csf2.army.mil/

7.18. Defense Enrollment Eligibility Reporting System (DEERS). The Defense Enrollment Eligibility Reporting System (DEERS) is a worldwide database of uniform services members (sponsors), their family members, and others who are eligible for military benefits. DEERS is used in the Real-Time Automated Personnel Identification System (RAPIDS). The RAPIDS is United States Department of Defense system that is used to issue the definitive credential within DOD for obtaining Common Access Card tokens in the DOD PKI. More information on the DEERS program can be found at https://www.dmdc.osd.mil/milconnect/faces/index.jspx?_af rLoop=5247895658679396&_afrWindowMode=0&_adf.ctr l-state=te5tzvn42_14

7.19. Equal Opportunity Program. The U.S. Army will provide EO and fair treatment for military personnel and

Family members without regard to race, color, gender, religion, national origin, and provide an environment free of unlawful discrimination and offensive behavior. Discrimination has no place in society and is detrimental to Army readiness. Leaders create and sustain effective units by eliminating discriminatory behaviors or practices that undermine teamwork, mutual respect, loyalty, and shared sacrifice of the men and women of America's Army. The Equal Opportunity (EO) program formulates, directs, and sustains a comprehensive effort to maximize human potential and to ensure fair treatment for all persons based solely on merit, fitness, and capability in support of readiness. EO philosophy is based on fairness, justice, and equity. More information on the Army EO program can be found at **http://www.apd.army.mil/pdffiles/r600_20.pdf.**

7.20. Exceptional Family Member Program. The Exceptional Family Member Program (EFMP) is a mandatory enrollment program that works with other military and civilian agencies to provide comprehensive and coordinated community support, housing, educational, medical and personnel services to Families with special needs. More information on the EFMP program can be found at **http://www.armymwr.com/family/efmp.aspx**

7.21. Financial Readiness Program. The Army Financial Readiness Program (FRP) provides a variety of education and counseling services to assist Soldiers and Families by increasing personal readiness and reducing financial stressors. Services include life-cycle education, personal financial training, advanced individual training, online financial readiness training and financial literacy gaming. The program provides financial guidance and support to Soldiers and their Families in the areas of general pay and allowances, entitlements, relocation, and credit reports. More information about FRP can be found at **http://www.myarmyonesource.com/default.aspx.** or **http://myarmybenefits.us.army.mil/Home/Benefit_Library /Federal_Benefits_Page/Financial_Readiness.html?serv=2 28**

    a. Having knowledge of what impacts your credit and the ability to obligate debt without becoming indebted will have a significant impact on maintaining individual Soldier readiness. Indebtedness has negative impacts on Soldier morale, personal and family security, and peace of mind. Personal financial management entails maintaining good credit and building financial growth.

b. Building personal financial growth. The best personal financial manager is the individual. The first and best rule to personal financial growth is to develop and adhere to a sound budget philosophy. If a Soldier does not really understand how to use a budget effectively, they will not be able to manage financial growth. Managing a budget is about assets and liabilities. A bank account is an asset; a car loan is a liability.

7.22. The First Sergeant Barracks Program. The purpose of the First Sergeant Barracks Program (FSBP) is to Provide the best quality unaccompanied housing in order to preserve and enhance the All-Volunteer Force and reinforce our commitment to providing a quality of life commensurate with their service and sacrifice to the nation. More information on the FSBP can be found at **http://www.imcom.army.mil/Organization/G4Facilitiesand Logistics/FirstSergeantsBarracksProgram.aspx**

7.23. Fort Family Outreach and Support Center. Fort Family Outreach and Support Center provides relevant and responsive information to support Army Reserve Soldiers and Families, and provides a single gateway to responsive Family Crisis Assistance, which is available 24 hours a day, 365 days per year. More information can be found at **https://www.arfp.org/fortfamily.html**

7.24. The Inspector General's Officer. The Inspector General's Office primary function is to ensure the combat readiness of subordinate units in their command. They investigate noncriminal allegations and some specific criminal investigations, but they help correct problems that affect the productivity, mission accomplishment, and morale of assigned personnel, which is vital to unit readiness. The IG provides assistance with inspections and compliance programs as well as teaching and training provided to units and their leaders.

The IG Office is a great resource to seek information and assistance when handling Soldier issues.

7.25. Military & Family Life Counseling Programs. Military & Family Life Counseling Program's (MFLC) licensed clinical providers assist Service Members and their Families with issues they may face throughout the cycle of deployment - from leaving their loved ones and possibly living and working in harm's way to reintegrating with their family and community. More information on the MFLC program can be found at **https://www.mhngs.com/app/programsandservices/mflc_p rogram.content**

7.26. Morale, Welfare, Recreation and Family Programs. Morale, Welfare and Recreation (MWR) and Family Morale, Welfare, and Recreation (FMWR) provides programs and services supporting Soldiers, Families, and civilians that promote resiliency and strengthen our Army. Services include Child, Youth, and School Services, Army Family Programs, Soldier Programs and Community Recreation, Family and MWR Business Initiatives, Armed Forces Recreation Centers, and MWR recreation delivery to theater operations. More information on MWR programs can be found at **http://www.armymwr.com/**

7.27. Private Public Partnerships. The Private Public Partnership Initiative (P3I) builds and enhances mutually beneficial partnerships between the civilian and military communities. This is accomplished by developing a mutually supportive environment for Soldiers, Veterans, and Family members striving to create and enhance career and training opportunities. More information on P3I can be found at **http://www.usar.army.mil/resources/Pages/Employer-Partnership-opportunities-and-Information.aspx**

7.28. Soldier for Life Program. Soldier for Life assist Soldiers in achieving the right mindset, obtain the necessary training and qualifications, and make the necessary connections through the Army, governmental and community efforts to successfully reintegrate Soldiers, veterans and their Families into civilian life. The Soldier for Life initiative focuses on a Soldier's lifecycle; that once a Soldier, always a Soldier. The four points highlight that a Soldier starts strong, serves strong, reintegrates strong and remains strong. When Soldiers are better reintegrated, they stay Army Strong, instilled in values, ethos, and leadership within communities. More information on the Soldier for Life Program can be found at **http://www.army.mil/Soldierforlife/**

7.29. Sexual Harassment/Assault Response and Prevention (SHARP). Sexual harassment and sexual assault violate everything the U.S. Army stands for including our Army Values and Warrior Ethos. The Army is aggressively addressing sexual assaults by first focusing on prevention through education and training. Army leaders encourage reporting and work hard to reduce the stigma associated with sexual violence. Once reported, the Army focuses on care for victims and thorough investigations and prosecutions to hold offenders accountable. The Army continually assesses the effectiveness of its sexual harassment/assault response and prevention efforts to ensure the Army is meeting the needs of the Soldiers, Department of the Army Civilians, family members and the nation. The Sexual Harassment/Assault Response and Prevention (SHARP) Programs mission is to reduce with an aim toward eliminating sexual offenses within the Army through cultural change, prevention, intervention, investigation, accountability, advocacy/response, assessment, and training to sustain the All-Volunteer Force. More information on SHARP can be found at **http://www.preventsexualassault.army.mil/index.cfm** or **http://www.sexualassault.army.mil/index_saam_2014.cfm**

7.30. Soldier For Life – Transition Assistance Program. The Soldier For Life – Transition Assistance Program (SFL-TAP) is a centrally funded and administered program that provides transition and job assistance services on major installations. It is a component of Soldier for Life transition and integral to Soldier development and support. SFL-TAP is worthy of every Leader's support and provides notification correspondence, registering for services, and workshops. The SFL-TAP Call Center can be reached by calling 1-800-325-4715. More information on the AFL-TAP program can be found at **https://www.acap.army.mil**

7.31. Total Army Sponsorship Program (TASP). The TASP assist Soldiers, civilian employees, and Families during the reassignment process. It assists Families geographically separated from the Soldier or civilian employee sponsor because of duty requirements. It improves unit or organizational cohesion and readiness by decreasing distractions that hamper personal performance and mission accomplishment, specifically by providing support and assistance, teaching teamwork, and encouraging development of a sense of responsibility. It supports the army's personnel life-cycle function of sustainment. More information on TASP can be found at http://www.militaryonesource.mil/moving?content_id=266791

7.32. Total Army Strong Program. Total Army Strong reaffirms the Army's commitment to the total Army family, builds trust and faith between the Army and its most precious resource, the people, and sets the foundation for a balanced system of programs and services. These programs and services will meet the unique demands of military life, foster life skills, strengthen and sustain physical and mental fitness and resilience, and promote a strong, ready, and resilient Army. This program succeeds Army Family Covenant and more information can be found at https://www.myarmyonesource.com/communitiesandmarketplace/totalarmystrong/totalarmystrong.aspx

This page intentionally left blank.

CHAPTER 8

LEADER TOOLS

Chapter 8

Leader Tools.

8.0. Ref.

- AR 1-201, Army Inspection Policy
- AR 600-20, Army Command Policy
- AR 600-25, Salutes, Honors, and Visits of Courtesy
- DA Pam 600-20, A Guide to Protocol and Etiquette
- TC 3-21.5, Drill and Ceremonies

8.1. Leadership Philosophy. Military organizations have mission statements that assist in making operational what is in the organization's vision statement. A philosophy is intended to articulate an individual's priorities within the context of the vision and mission statements. A Leadership philosophy lets people know what you expect, what you value and how you'll act. Leadership philosophies help:

- Keep the leader on course
- Let Soldiers know what the leader wants
- Provide clear leader intent
- Establish clear priorities
- Provide consistency that enhances trust and confidence
- Staff and Soldiers understand the leader's inner thoughts, beliefs, and expectations
- Communicate a vision and purpose to an organization

8.2. Army Training Network. The Army Training Network provides services to support training management issues. The ATN provides an online self-help service available such as UTM, ADP 7-0, METL Development, links to supporting doctrine, references, and training products. Training is one of a leaders three roles and ATN provides substantial support. Support services include:

- Training Solutions
- How to develop good trainers

- Training management best practices
- Unit provided examples
- DTMS tutorials Mall
- Links to other training resources.

NOTE: More information about ATN can be found at: **http://usacac.army.mil/cac2/atn/**

8.3. Army Career Tracker (ACT). The Army Career Tracker is an individual career management system aimed at supporting the lifecycle of the Soldier. The ACT encourages Soldier and their leaders to define career goals, create and ensure timelines are met for those goals and help fulfill objectives both inside and outside the Army. The ACT continues to improve on its capabilities to resource and inform Soldiers and leaders of initiatives and career development programs. The ACT provides the following features to assist both the Soldier and leader:

a. Encourages Soldiers to develop an individual development plan that tracks training, military education, civilian education, and a host of other development paths.

b. Access to SSD enrollment.

c. Automated sponsorship which standardizes procedures for requesting for a sponsor; management of the linkage between Soldier and Sponsor by the losing and gaining commands.

d. Provides links to other support sites such as MyPay.

e. Consolidates information from several systems and presents it at one central site.

f. It integrates Total Army Database, GoArmy Education, the Army Learning Management System, and the Army Training Requirements and Resources System (ATRRS)

Note: The following link provides access to the ACT website: **https://actnow.army.mil/**

8.4. Army Doctrine Publications (ADP).

a. The Department of the Army Publication that contains the fundamental principles of which the military forces or elements thereof guide their actions in support on National objectives. The Army has streamlined doctrine under the concept of Doctrine 2015. Publications are now broken down into four categories; Army Doctrine Publications (ADP), Army Doctrine Reference Publications (ADRP), Field Manuals (FM), Army Techniques Publications (ATP).

(1) ADPs are generally limited to approximately 10 pages and explains the fundamentals of the subject and how these support ADP 3-0, Operations. ADP consists of 16 publications. ADP 1 (The Army), 3-0 (Unified Land Operations, 6-22 (Army Leadership), and 7-0 (Training Units and Developing Leaders) are approved by the Chief of Staff of the Army. The Combined Arms Center (CAC) CG approves all other ADPs.

(2) For each ADP, there is an ADRP that provides detailed explanation of all doctrinal principles, which provide the foundational understanding so everyone in the Army can interpret it the same way. ADRP are publications of less than 100 pages and are approved by the CAC CG.

(3) There are currently 50 field manuals that contain TACTICS and PROCEDURES. The main body of an FM contains a maximum of 200 pages and describes how the Army executes operations described in ADPs. The CAC CG as the TRADOC proponent for Army Doctrine approves FM.

(4) ATP is authenticated versions of APD and contains TECHNIQUES. ATP have draft version on a WIKI site. Each technique publication has an assigned proponent responsible for monitoring input via WIKI and making changes to the authenticated publication. The approval authority is the proponent.

8.5. Travel Risk Planning System (TRiPS). US Army Combat Readiness Center. TRiPS is a risk mitigation program for individual planning to travel by their privately owned vehicles and uses an individual travel assessment to apply risk management controls if needed. For additional information reference **https://safety.army.mil/HOME.aspx**

8.6. Army Leader Book. The leader book is a tool maintained by leaders at all levels for recording and tracking Soldier proficiency on mission-oriented tasks. The exact composition of leader books varies depending on the mission and type of unit. Specific uses for the leader book are to–

- Track and evaluate Soldier's training status and proficiency on essential Soldier tasks.
- Provide administrative input to the chain of command on the proficiency of the unit for example; platoon, section, squad, team, or crew.
- Conduct Soldier performance counseling.

Note: Unit Leader Book examples are located at: https://atn.army.mil/dsp_template.aspx?dpID=450

a. **Organization:** The organization of the leader book is up to each individual leader. To be effective they must be well organized and "user friendly." Only essential training information is included in the leader book. The following is a recommended format that is applicable to all types of units with minor modifications:

- SECTION 1: Administrative Soldier data.
- SECTION 2: Company METL/platoon supporting collective task list with assessments.
- SECTION 3: CTT proficiency (survival skills).
- SECTION 4: Essential Soldier task proficiency and status.
- SECTION 5: Unit collective task proficiency.

This page intentionally left blank.

APPENDIXES

Appendix A.

Professional Reading List

Self-Development and personal growth are essential to developing leaders. Reading is an essential element of professional development. The scope and breadth of leader relevant issues in professional readings will serve to broaden and to deepen our understanding of our roles in leadership. The recommend list is not all-inclusive and will evolve as relevant leadership issues emerge. As a professional, it is important to create a personal course to read, study, reflect and apply in order to improve your understanding of our profession.

Band of Brothers: E Company, 506th Regiment, 101st Airborne from Normandy to Hitler's Eagle's Nest
Stephen E. Ambrose // New York: Simon & Schuster, 2001

Constitution of the United States
Available online at
http://www.archives.gov/nationalarchives-experience/charters/constitution.html

Gettysburg Address
Available online at
http://www.archives.gov/nationalarchives-experience/charters/constitution.html

The Profession of Arms
John Winthrop Hackett // New York: Macmillan, 1983

We Were Soldiers Once ... and Young: La Drang—the Battle That Changed the War in Vietnam
Harold G. Moore and Joseph L. Galloway // New York: Harper Torch, 2002

The Starfish and the Spider: The Unstoppable Power of Leaderless Organizations

Ori Brafman and Rod Beckstrom // New York: Penguin Group, 2006

Surviving the Shadows: A Journey of Hope into Post-Traumatic Stress
Bob Delaney with Dave Scheiber, Amazon/Barnes and Noble

The Servant; A Simple Story About The True Essence of Leadership
Jim Hunter, Amazon/Kindle

The Richest Man in Babylon
George S. Clason, Penguin books, 1926

Up Front
Bill Mauldin, WW. Norton & Company, 2000

The 16 Sixteen-Personality Types, Descriptions for Self-Discovery
Linda V. Berens and Dario Nardi,   Radiance House, 1998.

George C. Marshall, Soldier-Statesman of the American Century
Mark A. Stoler, Twayne Publishers, Simon   and Schuster MacMillian, New York 1989

Appendix B

References

REQUIRED PUBLICATIONS
Most Army doctrinal publications are available online:
http://www.apd.army.mil/.
Most joint publications are available online:
http://www.dtic.mil/doctrine/new_pubs/jointpub.htm

ADP 1. *The Army.* 17 September 2012.

ADP 3-0. *Unified Land Operations.* 10 October 2011.

ADP 6-0. *Mission Command.* 17 May 2014.

ADP 6-22. *Army Leadership.* 1 August 2012.

ADP 7-0. *Training Units and Leader Development.* 23 August 2012.

ADRP 1. *The Army Profession.* 14 June 2013.

ADRP 1-02. *Terms and Military Symbols.* 2 February 2015

ADRP 3-0. *Unified Land Operations.* 16 May 2012.

ADRP 6-0. *Mission Command.* 17 May 2012.

ADRP 6-22. *Army Leadership.* 1 August 2012.

ADRP 7-0. *Training Units and Developing Leaders.* 23 August 2012.

ALARACT 126/2014. Transition Of Advanced Leader Course Common Core To Structured Self Development Level 2. 15 May 2014. Retrieved from http://www.signal.army.mil/images/SIGCoEOrgs/RNCOA/Documents/Documents/ALARACT_126-2014.pdf

AR 27-10. *Military Justice.* 03 October 2011.

AR 350–1. *Army Training and Leader Development.* 19
August 2014.

AR 600-20. *Army Command Policy.* 6 November 2014.

JP 1-02. *DoD Dictionary of Military and Associated Terms,* 15
December 2014.

MCM (2012). *Manual for Courts Martial 2012 Edition.* 5
April 2012. Retrieved from
http://www.apd.army.mil/pdffiles/mcm.pdf

TC 3–21.5. *Drill and Ceremonies.* 20 January 2012.

U.S. Department of War. *Abstract of Infantry Tactics.* 1829.
Retrieved from
http://www.carlisle.army.mil/ahec/research.cfm

RELATED PUBLICATIONS
Most Army doctrinal publications are available online:
http://www.apd.army.mil/.

Adams, Alison Dr.,"Self-Esteem" *Ezine,* April 2012, Issue 28.

AR 1-201. *Army Inspection Policy.* 04 April 2008.

AR 190-47. *The Army Corrections System.* 15 June 2006.

AR 215-1. *Military Morale, Welfare, and Recreation
Programs and Non-appropriated Fund
Instrumentalities.* 24 September 2010.

AR 385-10. *The Army Safety Program.* 27 November 2013.

AR 600-8-8. *The Total Army Sponsorship Program.* 04 April
2006.

AR 600–9. *The Army Body Composition Program.* 28 June
2013.

AR 600-25. Salutes, Honors, and Visits of Courtesy. 24 October 2004

AR 600-85. *The Army Substance Abuse Program.* 28 December 2012.

AR 600-100. *Army Leadership.* 08 March 2007.

AR 601–280. *Army Retention Program.* 31 January 2006.

AR 608-1. *Army Community Service.* 12 March 2013.

AR 608-18. *The Army Family Advocacy Program.* 30 October 2007.

AR 608-75. *Exceptional Family Member Program.* 22 November 2006.

AR 621-5. *Army Continuing Education System.* 11 July 2006.

AR 623–3. *Evaluation Reporting System.* 31 March 2014.

AR 670–1. *Wear and Appearance of Army Uniforms and Insignia.* 15 September 2014.

AR 735-5. *Property Accountability Policies.* 10 May 2013.

AR 930-4. *Army Emergency Relief.* 22 February 2008.

AR 930-5. *American National Red Cross Service Program and Army Utilization.* 1 February 2005.

Arms, L.R., *A History of the NCO,* US Army Museum of the Noncommissioned Officer. March 2007.

Army Directive 2011-19, *Expedited Transfer or Reassignment Procedures for Victims of Sexual Assault.* 3 October 2011.

ATP 6-22.1. *The Counseling Process.* 1 July 2014.

Center for Army Leadership, *Leader Development Improvement Guide*, November 2014.

Center for the Army Profession and Ethic, *The Army Ethic White Paper*, 11 July 2014.

Center for the Army Profession and Ethic. *Stand Strong: Senior Leader Guide.* 11 December 2013

Chandler, Raymond F., III, SMA, "The Profession of Arms and the Professional Noncommissioned Officer," *Military Review - The Profession of Arms Special Edition* (Sep 2011):

DA PAM 350-58. *Army Leader Development Program.* 08 March 2013.

DA PAM 385-10. *Army Safety Program.* 23 May 2008.

DA PAM 600-20. *Personnel, General – Junior Officer Retention.* 1 August 1969.

DA PAM 611-21. *Military Occupational Classification and Structure.* 22 January 2007.

Elder, Dan CSM(R) and Sanchez, Felix CSM(R) "The History of the NCO Creed," *NCO Journal*, Summer 1998.

FM 6-0. *Commander and Staff Organization and Operations.* 5 May 2014.

FM 6-22. *Army Leadership: Competent, Confident, and Agile.* 12 October 2006.

FM 7-15. *The Army Universal Task List.* 27 February 2009.

FM 27-10. *The Law of Land Warfare.* 18 July 1956.

Lesperance, David A. COL(R), Developing Leaders for Army 2020, U.S. Army War College, Class 2012

Moss, James A., *Noncommissioned Officers' Manual*. 1917. Retrieved from http://www.carlisle.army.mil/ahec/research.cfm

Nickerson, Thomas COL(R), "The Making of Army Strong". November 8, 2006. Retrieved from http://www.army.mil/article/568/the-making-of-army-strong/.

Stone, David R. *A Military History of Russia*, 2006.

Title 10 USC 1331. *Retired Pay for Non-Regular Service*. 2006 Edition.

US Congress, "Our Flag", *Senate Document 105-013*, 5 November 1997.

USAR 608-1. *Army Reserve Family Programs.* 13 March 2013.

RECOMMENDED READING

AR 350-10. *Management of Army Individual Training Requirements and Resources.* 03 September 2009.

AR 600-4. *Remission or Cancellation of Indebtedness.* 07 December 2007.

AR 600–8. *Military Human Resources Management.* 11 April 2014.

AR 600-8-2. *Suspension of Favorable Personnel Actions (FLAG).* 23 October 2012.

AR-600-8-11. *Reassignment.* 01 May 2007.

AR 600–8–19. *Enlisted Promotions and Reductions.* 30 April 2010.

AR 621-202. *Army Educational Incentives and Entitlements.* 03 February 1992.

AR 635-200. *Active Duty Enlisted Administrative Separations.* 06 June 2005.

AR 870-5. *Military History: Responsibilities, Policies and Procedures.* 21 September 2007.

DA PAM 385-1. *Small Unit Safety Officer/NCO Guide.* 23 May 2013.

DA PAM 600-25. *U.S. Army NCO Professional Development Guide.* 28 July 2008.

FM 7-22. *Army Physical Readiness Training.* 26 October 2012.

PRESCRIBED FORMS
None

REFERENCED FORMS
Unless otherwise indicated, DA Forms are available on the Army Publishing Directorate (APD) web site: http://www.apd.army.mil/

DA Form 2028. *Recommended Changes To Publications And Blank Forms.*

DA Form 2627. *Record of Proceedings Under Article 15, UCMJ.*

WEBSITE
Photo Images where retrieved from ARMY.MIL, The Official Homepage of the United States Army. http://www.army.mil/.

Army Reserve Family Programs, http://www.arfp.org/.

CAPE Website – Center for the Army Profession and Ethic, http://cape.army.mil

Institute of Heraldry, U.S. Army Flag and Streamers-Flag Information, http://www.tioh.hqda.pentagon.mil/.

NCO Creed-United States Army,
   http://www.army.mil/values/nco.html.

Structured Self Development,
   https://usasma.bliss.army.mil/page.asp?id=9.

Title 10 USC 1331,
   http://www.gpo.gov/fdsys/granule/USCODE-2011-
   title10/USCODE-2011-title10-subtitleA-partII-
   chap67-sec1331.

Appendix C.

Glossary/Acronyms/Abbreviations and Terms

AAR	after action review
ADP	Army doctrine publication
ADRP	Army doctrine reference publication
AR	Army regulation
DOD	Department of Defense
FM	field manual
JP	joint publication
MSAF	Multi-Source Assessment and Feedback
NATO	North Atlantic Treaty Organization
NCO	NCO
SHAEF	Supreme Headquarters Allied Expeditionary Force
TC	training circular
TTP	tactics, techniques, and procedures
Army Professional	A member of the Army Profession who meets the Army's certification criteria of competence, character, and commitment.
Certification	Verification and validation of an Army Professional's Character, Competence, and Commitment to fulfill responsibilities and

	successfully perform assigned duty with discipline and to standard.
Character	Dedication and adherence to the Army Ethic, including Army Values, as consistently and faithfully demonstrated in decisions and actions.
Command	The authority that a commander in the armed forces lawfully exercises over subordinates by virtue of rank or assignment. Command includes the authority and responsibility for effectively using available resources and for planning the employment of, organizing, directing, coordinating, and controlling military forces for the accomplishment of assigned missions. It also includes responsibility for health, welfare, morale, and discipline of assigned personnel. (JP 1-02)
Commitment	Resolve to contribute Honorable Service to the Nation and accomplish the mission despite adversity, obstacles, and challenge.
Competence	Demonstrated ability to successfully perform duties with discipline and to standard.
Leader development	Leader development is a deliberate, continuous, sequential, and progressive process grounded in the Army Values. It grows Soldiers and Army Civilians into competent, confident leaders capable of directing teams and organizations. (AR 350-1)

Leadership	The process of influencing people by providing purpose, direction, and motivation to accomplish the mission and improve the organization. (ADP 6-22)
Mentorship	The voluntary developmental relationship that exists between a person of greater experience and a person of lesser experience that is characterized by mutual trust and respect. (AR 600-100)
Mission command	The exercise of authority and direction by the commander using mission orders to enable disciplined initiative within the commander's intent to empower agile and adaptive leaders in the conduct of decisive action. (ADP 6-0)